Principles of
Organic Stereochemistry

Studies in Organic Chemistry

Executive Editor

Paul G. Gassman

*Professor and Chairman
Department of Chemistry
University of Minnesota
Minneapolis, Minnesota*

Other Volumes in Preparation

Principles of
Organic Stereochemistry

BERNARD TESTA

Department of Medicinal Chemistry
School of Pharmacy
University of Lausanne
Lausanne, Switzerland

MARCEL DEKKER, INC.   New York and Basel

Library of Congress Cataloging in Publication Data

Testa, Bernard.
    Principles of organic stereochemistry.

    (Studies in organic chemistry ; v. 6)
    Bibliography: p.
    Includes indexes.
    1. Stereochemistry.   2. Chemistry, Physical
organic.   I. Title.
QD481. T42   547'. 1'223   78-10138
ISBN 0-8247-6758-6

MARCEL DEKKER, INC.
270 Madison Avenue, New York, New York 10016

Current printing (last digit):
10 9 8 7 6 5 4 3

PRINTED IN THE UNITED STATES OF AMERICA

Dedicated

in part to such Oriental fairies as Darjeeling, Oolong,
Gunpowder, Souchong, Keemun, Chun Me,
for delectable nocturnal support

and fully to Jacqueline,
for what she is, gives, and shares

Foreword

An essential prelude to the development of the field of stereochemistry was the discovery by Biot that solutions of many organic compounds had the property of rotating the plane of polarized light passing through them. These observations were published in 1815–1817. Not much could be made of them, however, until the great French chemist Pasteur recognized that optical activity at the molecular level was caused by an asymmetric grouping of atoms in the molecule, and that enantiomers had equal and opposite effects upon polarized light (1848–1860). How these earlier ideas related to the actual three-dimensional structures and properties of molecules then followed from the publications in 1874 by J. H. van't Hoff and J. A. Le Bel. Since that time the subject of stereochemistry has developed and expanded immensely.

It is convenient to divide stereochemistry into two parts, static and dynamic. Static stereochemistry may be considered as a study of the properties of molecules in their ground states. Dynamic stereochemistry can, conversely, be considered to be a study of the properties which result from the molecule moving from one state to another. Dynamic stereochemistry therefore covers the area of chemical reactions (where the molecule moves from the ground state to a transition state), as well as cases in which no reaction occurs, but the molecule moves between two states. Variable temperature NMR spectroscopy is the best known example.

The present text is a very readable account, concerned in the main with static stereochemistry. The dynamic aspects of the subject are generally covered in writings on organic reactions and in the more specialized literature, and are treated here only briefly.

The level of the present work is suitable for an advanced undergraduate or a beginning graduate student in chemistry. It is also a convenient reference source for the chemist whose specialty is elsewhere. Many good books and chapters on the general subject exist. Most of these are either rather old, or else quite specialized and limited to a small area of the overall subject. Some have become dated through the passage of time and with the invention or discovery of new tools for studying stereochemistry. New examples and different kinds of systems have been discovered in recent years, and new kinds of nomenclature have been developed to deal with situations not previously considered.

The present work examines stereochemistry in a fully modern and up-to-date way. A perusal of the Contents will outline this coverage for the reader, and need not be repeated here in detail. It might be mentioned that a few of the chapters are of special significance and cover subjects in a detail not usually met with in other texts. These include Chapter 3, on bonding and the forces of which observed stereochemical features are consequences. Nomenclature is covered fully and lucidly, especially in Chapter 4. Especially interesting is Chapter 11, on pro-stereoisomerism. This is a subject, the importance of which has been only recently recognized, but which is clearly fundamental at the molecular level to almost all life processes.

Norman L. Allinger

Professor,
Department of Chemistry
University of Georgia
Athens, Georgia

Preface

If the human creatures will not understand Relativity,
very well; but they must understand Relatedness. *

The goal of chemistry is to study and describe the properties of matter
at a particular state of aggregation and organization, namely, that of molecules. Like all other sciences, it is in a permanent state of evolution;
models and theories stand as milestones on this endless road.

The history of chemistry intermingles with the conquest of dimensions,
with the increase in the dimensionality of its models. In the early historical stages of chemistry, some scientists believed molecules to have no
physical structure and to be merely useful abstractions. Other chemists,
while accepting the physical reality of molecules, had nothing more than
rudimentary symbols to use as models.† These were the low-dimensionality
stages of chemistry.

Two-dimensional chemistry is still widely practiced today with pen and
paper, the reason being that this simple visualization of molecules is sufficient to account for many facts. However, the geometrical conception of
molecules in the three dimensions of space is not new. Among others, the
fathers of stereochemistry are Louis Pasteur with his experimental work
carried out around 1850, and van't Hoff and Le Bel who independently proposed in 1874 the model of the tetrahedral carbon as a basis of molecular
dissymmetry and optical rotation.†

---

*Ursula K. Le Guin, "Direction of the Road," in The Wind's Twelve
Quarters, Bantam, New York, 1976.
†F. G. Riddell and M. J. T. Robinson, Tetrahedron 30:2001 (1974).

The three-dimensional model of chemistry, namely, stereochemistry in a narrow sense, is a static and frozen conception of molecules. A temporal dimension is lacking, which allows us to depict inter- and intra-molecular rate processes. The integration of this fourth dimension into chemistry and more specifically into stereochemistry has progressed during the last decades to generate dynamic stereochemistry. In fact, stereochemistry must be dynamic since such concepts as isomers and isomerization processes, for example, are inseparable. This point will be considered again later.

The three spatial dimensions and the temporal dimension just discussed do not correlate among themselves. Geometrically speaking, they are orthogonal. This would not be the case with an energy dimension. The latter indeed correlates to some degree with the temporal dimension and the two terms cannot be included together in the same model without difficulty.

The recent years have allowed significant advances in our visualization of molecules, in our understanding of their properties, and in our models. One likes to consider electronic terms as the additional dimension(s) of current chemical models, and electronic properties as the parameters occupying these dimensions. Although electronic dimensions do somewhat correlate with the four "basic" dimensions, intuitively this correlation is far too small to forbid the use of models including all these terms. Indeed, theoretical studies of stereoelectronic control in dynamic processes have resulted in many-dimensional models.[*]

As we gain insight into molecules and their properties, so our chemical models increase their dimensionality and consequently their complexity. In the present book, I consider mainly three-dimensional chemistry, that is to say molecules in space. More specifically, the book is first aimed at discussing and classifying relationships between the three-dimensional structure of molecules. However, a study of relatedness would not be complete and fruitful without considering processes of interconversion and factors influencing them. This is why the temporal dimension is ubiquitous in the text, and electronic factors are mentioned when necessary. Energies and mechanisms of interconversions are also considered, but only as additional aspects of structural relatedness. It is the latter which provides the backbone of the book and its classification of matters.

Conceptual stereochemistry has greatly progressed in recent years, and I have attempted to include all major advances. The new stereochemical nomenclatures have now become classical and are presented in some detail. Recently crystallized concepts such as prostereoisomerism or the classification of steric relationships receive due consideration in the light of

---

[*] N. D. Epiotis, W. R. Cherry, S. Shaik, R. L. Yates, and F. Bernardi, Structural Theory of Organic Chemistry, Springer-Verlag, Berlin, 1977.

their considerable utility and even greater promise. Much is also said about conformations and factors influencing them; I believe that the significance of conformation in chemistry, biochemistry, and medicinal chemistry cannot be overestimated, and this belief certainly shows through the length of the chapters dealing with conformational aspects.

The present work is aimed at bringing to the student a clear even if necessarily incomplete view of stereochemistry, at refreshing and clarifying the memory of the professional chemist, and at allowing every reader to go beyond the level of the text by making use of the many references cited.

I owe gratitude to several people for help or advice. Professor William P. Purcell read an early version of the manuscript and made useful comments. The text has gained much from the careful reading and highly constructive comments of Professor Hans Wynberg. The renowned efficiency of my wife Jacqueline and her support have helped me tremendously at every stage of preparation of the book, while Philippe and Christine contributed with their unfailing and catching liveliness. No thanks are due for the artwork, it being my own hand's. If you enjoy looking at some of the diagrams, imagine that the daily vista of the Lake of Geneva may be aesthetically stimulating.

Bernard Testa

Contents

1

Symmetry Elements and Operations. Point Groups

Symmetry, in Greek, means "with measure." As such, the word conveys an idea of equilibrium and harmony, and has a clear aesthetic import. For scientists, it provides a way of rigorously describing certain geometrical properties of a given object. Symmetry, a mathematical tool and an aesthetic index, is a privileged means of uncovering some of the beauty in science, and it helps us appreciate the structural complexity and richness of those eye–invisible entities known as molecules.

In order to specify the symmetry class of molecules, a shorthand notation is used. The type of symmetry to which a molecule (or any object) belongs is known as its point group. The point group of any molecule "A" is the ensemble of the symmetry operations which transform A into a molecule to which it is superimposable. These symmetry operations are based on symmetry elements, and both terms are necessary in order to define symmetry.

1.1 Symmetry Elements and Symmetry Operations

Comprehension of symmetry operations is required when defining symmetry elements, and vice versa. Therefore, the two terms lack independent meaning and must be considered together.

Symmetry elements

Proper (or simple) axes of rotation (C_n)
Planes of symmetry (σ)
Centers of symmetry (i) (or inversion centers)
Rotation-reflection axes (S_n) (also called mirror axes, improper axes, or alternating axes)

1

Symmetry operations

Rotations (C_n)
Reflections (σ)
Inversions (i)
Rotation-reflections (S_n)

The pseudo-operation of identity is not considered here.

A molecule is said to have a <u>symmetry axis</u> (C_n) of order n (n-fold axis of symmetry) if a <u>rotation</u> of 360°/n around this axis yields an arrangement which cannot be distinguished from the original. For example, the molecule of water (I) has a twofold axis of symmetry (C_2), and chloroform (II) has a C_3 axis. Benzene (III) has a C_6 axis perpendicular to the plane of the molecule and passing through the geometric center, and six additional C_2 axes lying in the molecular plane. In this example, C_6 is the axis having the higher order, and becomes the principal axis. An extreme case is represented by linear molecules such as acetylene (IV) which have a $C_\circ$ axis, since even an infinitesimal rotation (360°/∞) about this axis results in an orientation indiscernible from the original. The other extreme case, the trivial onefold axis C_1, is never considered because all molecules possess an infinite number of C_1 axes.

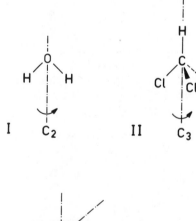

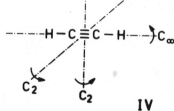

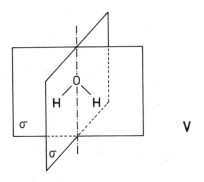

When a plane divides a molecule into two symmetrical halves, it is called a plane of symmetry (σ). By definition, σ is a mirror plane passing through the molecule in such a way that the reflection of all atoms through the plane yields a three-dimensional arrangement indistinguishable from the original. In a molecule with a plane of symmetry, all atoms out of the plane exist in pairs. The molecule of H_2O (V) possesses two planes of symmetry which intersect along C_2, while chloroform has three such planes intersecting along C_3 and each including H-C-Cl.

All planar molecules have at least one plane of symmetry, identical with the molecular plane; linear molecules possess an infinite number of σ planes, intersecting along C_∞. Planes of symmetry perpendicular to the principal axis are labeled σ_h (h = horizontal), while those containing the principal axis are marked σ_v (v = vertical).

A center of symmetry (i) exists for a molecule if every atom of the molecule has a symmetrical counterpart with respect to this center. In such a molecule, inversion of all atoms relatively to the center of symmetry again results in a three-dimensional structure indistinguishable from the original. No more than one center of symmetry can exist per molecule. For benzene (III), the center is at the intercept of C_6 and of the 6 C_2.

The symmetry operation known as rotation-reflection (S_n) involves two manipulations, namely, rotation of 360°/n about an axis designated S_n, followed by reflection through a plane perpendicular to S_n and passing through the molecule (plane of symmetry). These operations have been considered separately as C_n and σ; their combination results in a distinct operation. Molecules possessing an improper axis S_n display reflection symmetry, i.e., they are superimposable on their reflection or mirror image. Consider, for instance, trans-dichloroethylene (VI), whose rotation of 180° around S_2 followed by reflection in a reflection plane restores the original orientation. The two operations have no priority over each other, and it is apparent that S_n exists when neither C_n nor σ are present. For convenience, a mirror plane (outside the molecule) is often used instead of a reflection plane, with identical results. Let us also consider

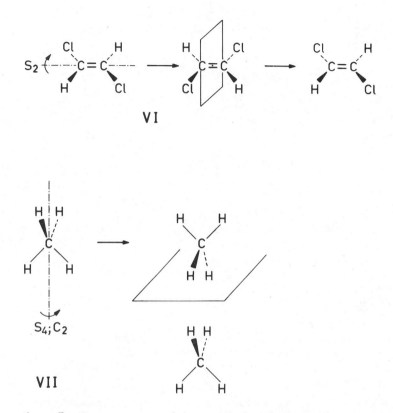

VI

VII

the reflection symmetry of the tetrahedral carbon (VII); it has three C_2, identical with three S_4. A 90° rotation about an S_4, followed or preceded by reflection in a mirror plane (or, properly, a reflection plane) restores the original orientation.

1.2 Point Groups

A group of symmetry is the ensemble of all symmetry operations which can convert a given molecule into orientations indistinguishable from the original. Therefore, it is the ensemble of all symmetry elements possessed by a given molecule. Although the number of molecules is almost infinite, the possible combinations of symmetry elements and operations are relatively few. These combinations are called point groups--they must leave a specific point of the molecule unchanged--in contrast to space groups, which are associated with operations applied to unit cells, e.g., translation, and which are beyond the scope of these pages.

Point groups can be classified into two main groups, structures lacking reflection symmetry and structures possessing reflection symmetry. The reader will soon realize that such a classification is fundamental for the comprehension of stereochemistry.

Molecular structures without reflection symmetry (no σ plane) are called dissymmetric or chiral structures. Chirality (from the Greek word "chier," meaning hand) is the property displayed by a molecular structure (or any object, e.g., a hand) which is nonsuperimposable on its mirror image; this is also referred to as "handedness." No chiral structure can have a σ plane. If a C_n (n > 1) is also absent, the structure lacks all symmetry elements and is called asymmetric (point group C_1). An asymmetric carbon atom (VIII) illustrates this point group.

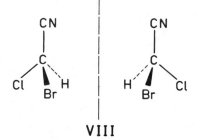

VIII

Molecules possessing axial symmetry (one or more C_n) can be dissymmetric but not asymmetric. Those having one symmetry axis only build the point groups C_n. Not uncommon are molecules belonging to point group C_2, i.e., having one C_2 as the only element of symmetry. Dichloroallene (IX) belongs to this point group: it is clearly chiral (IXa), and the C_2 symmetry is best seen using Newman projections (IXb).

Molecules having a principal C_n axis and n C_2 axes in a plane perpendicular to C_n are said to possess dihedral symmetry (point groups D_n). Of course, the plane containing the n C_2 axes must not be a σ plane. A three-bladed propeller has a C_3 axis passing through its center and three C_2 axes along each blade; therefore, it has D_3 symmetry.

Molecular structures with reflection symmetry are called nondissymmetric or achiral, rather than the ambiguous term "symmetric." Molecules with only a σ plane (no C_n) belong to point group C_s, for example, a monosubstituted cyclopropane derivative such as structure X. Some examples of molecules having an S_n axis but no σ plane are known (point group S_n, n being even). Nevertheless, such molecules obviously have reflection symmetry and this shows that a σ plane is not a necessary condition for reflection symmetry. The spirane molecule, structure XI, has an S_4 axis coincident on a C_2 axis, but no σ plane. After a 90° rotation along S_4, the molecule is superimposable on its mirror image.

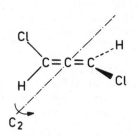
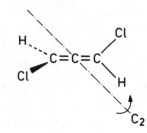

IXa

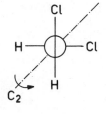

IXb

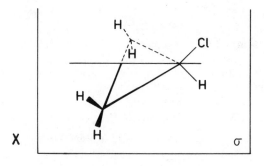

X

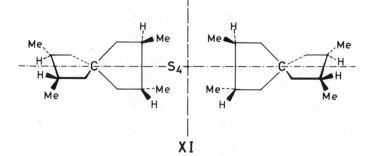

XI

In many cases, however, achiral molecules have both σ planes and C_n axes. With one C_n and n σ planes intercepting at C_n, the point groups are C_{nv} and the planes are σ_v. For example, the molecule of water (I, V) is C_{2v} and chloroform (II) is C_{3v}. Chloroacetylene has a C_∞ and $\infty\,\sigma_v$ like acetylene (IV), but no σ_h, it is of $C_\infty v$ symmetry.

Molecules with one C_n axis and one σ_h plane, but no σ_v planes, belong to groups C_{nh}. trans-1,2-Dichloroethylene (VI, XII) is such a case (C_{2h}). Molecules having one C_n axis and n C_2 axes (dihedral symmetry) with n σ_v planes, but no σ_h plane, belong to D_{nd} groups (d = diagonal). If

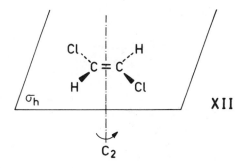

XII

TABLE 1 Principal Point Groups

Chiral groups		Achiral groups	
Type of group	Elements	Type of group	Elements
C_1	No symmetry element (asymmetric)	C_s	σ
		S_n	S_n (n even)
C_n	C_n (n > 1) (dissymmetric)	C_{nv}	C_n, n σ_v
D_n	C_n, n C_2 (dissymmetric)	C_{nh}	C_n, σ_h
		D_{nd}	C_n, n C_2, n σ_v
		D_{nh}	C_n, n C_2, n σ_v, σ_h
		T_d	4 C_3, 3 C_2, 6σ
		O_h	3 C_4, 4 C_3, 6 C_2, 9σ
		K_h	All symmetry elements

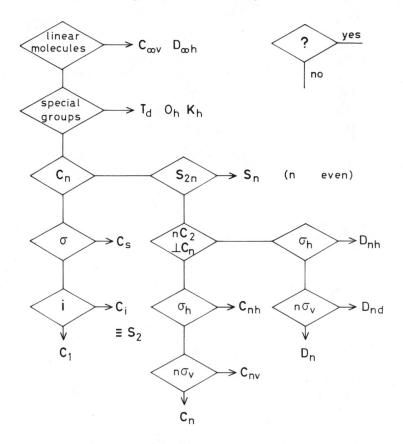

SCHEME 1 Scheme for the selection of point groups.

the σ_h plane is also present, then the point groups are D_{nh}. D_{nd} and D_{nh} are groups of higher symmetry than the previous groups. For example, benzene has D_{6h} symmetry (one C_6, six C_2, six σ_v, one σ_h), while acetylene (IV) has $D_{\infty h}$ symmetry (<u>cylindrical symmetry</u>).

Point groups of a higher order of symmetry have several C_n axes (n is larger than 2). They include group T_d, (<u>tetrahedral symmetry</u>), group O_h, (<u>octahedral symmetry</u>), and group K_h, (centrosymmetric group, <u>spherical symmetry</u>). For a summary of the above, see Table 1 and Scheme 1.

2

Molecular Representations

Chemistry, as an experimental and theoretical science, has developed its own language and symbolism. Chemists are hard at work to unravel the information contained in such entities as molecules, ions, and radicals. This information must then be stored with minimal loss for effective preservation, fast transmission, and easy theoretical handling, hence, the basis for chemical symbolisms as evidenced by several types of chemical representation.

Several models have been developed in order to represent the three-dimensional structure of molecules, as studied by stereochemistry and other branches of structural chemistry. These models are conventional representations and are of three types, namely, three-, two-, and one-dimensional.

Three-dimensional representations of molecules are called molecular models. They may or may not be scaled. Space-filling models use spheres, hemispheres, and any other convenient form, and represent the van der Waals volumes in a molecule. Skeletal models merely represent the framework of a molecule (nuclei and bonds). No further attention is given to molecular models in this study. The interested reader is referred to a thorough review [1], and to an illuminating discussion on their advantages and shortcomings by Mislow [2].

One-dimensional representations of molecules (e.g., Wiswesser line notation, WLN, Ref. 3) have been developed for the computer storage and handling of molecular structures. Relevant to the present context are stereochemical nomenclature rules, as used in the naming of molecules. Due attention will be given to these rules in the next chapters.

The remainder of the present chapter is devoted to the two-dimensional representation of spatial structures. The conventions to be presented and compared make use of diagrammatic symbols in order to depict as faithfully and completely as possible the stereochemical properties of molecules.

For instance, consider a carbon atom substituted by four ligands a, b, c, and d, which occupy the vertices of a tetrahedron (I). This tetrahedral model is rarely used, for the obvious reason that the chemical bonds are not apparent. It is useful, however, in introducing one of the most frequently used models, the flying-wedge representation (II, III). In such diagrams, bonds pointing toward the observer are symbolized by a wedge, bonds pointing away by a broken rather than a dotted line, and bonds in the plane of the paper by a continuous line. In the case of substituents adjacent to a cyclic structure, a heavy line usually replaces the wedge (IV). But for a bond linking two atoms in a chain or a cycle, a broken line is misleading because there is no way to know which of the two atoms is more remote from the observer. A wedge clearly conveys this information.

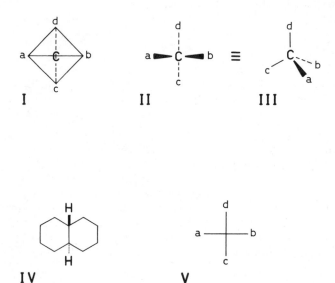

A simplification of the flying-wedge representation is the Fischer projection. All bonds are drawn as solid lines, with the understanding that horizontal bonds point toward the observer and vertical bonds point away. The central tetrahedral carbon atom is omitted. In Fischer projection, structure II becomes structure V but this projection requires that the flying-wedge representation first be drawn with the two wedges horizontal and the two dotted lines vertical. In other words, structure III has first to be viewed as structure II. It can be seen that the Fischer projection is not as illustrative as the flying-wedge representation. However, it has its value, for example, in the representation of the linear forms of monosaccharides.

The perspective representation is one of the most convenient ways of drawing molecules in space. Consider, for example, 1-chloro-2-fluoro-ethane (VI). The molecule has been taken as "frozen" in a given conformation (see Chapter 9), and structure VI conveys the information regarding the relative disposition in space of the various constitutive atoms. Perspective representations are noticeably useful for large molecules, in particular, cyclic systems. In the perspective structure drawing of cyclohexane (VII) and cis-decalin (VIII), the C-H bonds have been drawn, but the hydrogen atoms have been omitted. Quite often, the combined use of wedges, heavy lines, and perspective drawing, as shown for cyclohexane (IX) and adamantane (X), greatly increases the perspective effect, and, therefore, the impact of a structure drawing.

Just as the Fischer projection is a planar projection of the flying-wedge representation, so may the Newman representation be considered

VI

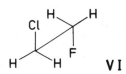

VII

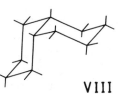

VIII

IX

X

as a planar projection of perspective representation. Here a molecule is viewed along the axis of a given bond, the adjacent groups being projected on a plane perpendicular to the reference axis. In the case of 1-chloro-2-fluoroethane (VI), viewing the molecule from C-1 along the C-C bond and using the conventional symbolism produces structure XI. Similarly, structure XII depicts one of the conformations of acetaldehyde.

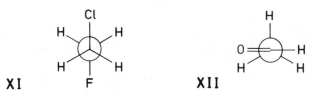

XI   **XII**

The value of Newman representations is evidenced by the fact that upon transformation of structure VI into structure XI, no information is lost regarding the topology of the six ligands. Another point of interest lies in the representation of alicyclic structures. Thus, cyclohexane, observed from the left-hand side in structure VII, can be represented as structure XIII. cis-Decalin, a more complex structure, is drawn as structure XIV. (In the latter case, the building of a skeletal molecular model greatly facilitates comprehension.)

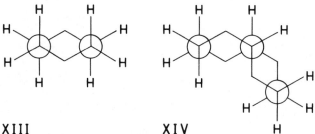

XIII   **XIV**

The perspective representation of molecules may in some cases be improved by the use of ball-and-stick drawings. Hydrogen and carbon atoms are represented by small and larger spheres, respectively. Heteroatoms are indicated by spheres bearing the symbol of the element. Ball-and-stick representations must be drawn from a skeletal molecular model, and, while not highly accurate, they are, nevertheless, of great value in presenting an easily understood perspective picture of a molecule. Norephedrine is represented in structure XV as an example. However, one increasingly sees X-ray diagrams presented in a comparable fashion. These diagrams are computer-drawn and certainly allude to a high degree of accuracy.

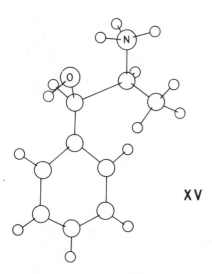

X V

The main interest of ball-and-stick representations is in stereoscopic drawings, one example of which is given in structure XVI. Stereoscopic views can be obtained by hand with a pair of stereoscopic line drawings. All published stereoscopic views, however, are computer-drawn. Each view in a stereoscopic pair represents the object as seen by the left or the right eye, respectively. By visually superimposing the left view seen by the left eye on the right view seen by the right eye, a stereoscopic image is obtained. This vision is called stereopsis, and is based upon an optical illusion. A simple stereoscopic viewer permits stereopsis, although this can also be attained by holding a piece of cardboard between the eyes. With

XVI

a little practice, such external help can be eliminated. It must be noted, however, that many people cannot achieve stereopsis. For further details, see Mills and Speakman [4], from which much of the above information is taken.

References

1. A. Walton, The use of models in stereochemistry, in Progress in Stereochemistry, vol. 4 (edited by B. J. Aylett and M. M. Harris), Butterworths, London, 1969, pp. 335-375.

2. K. Mislow, Introduction to Stereochemistry, Benjamin, New York, 1966, pp. 42-46.

3. E. G. Smith and P. A. Baker, The Wiswesser Line-Formula Chemical Notation (WLN), 3d ed., Chemical Management Inc., Cherry Hill, New Jersey, 1976.

4. H. H. Mills and J. C. Speakman, Crystallography and stereochemistry, in Progress in Stereochemistry, vol. 4 (edited by B. J. Aylett and M. M. Harris), Butterworths, London, 1969, pp. 273-298.

3

Electronic Bases of Molecular Structure

By the term molecular structure, we understand a general concept embodying all the elements which control the "shape" of a molecule in the three spatial and the temporal dimensions (see also the Preface). The stereochemistry of many molecules is, of course, a major component of their structure, which requires knowledge of the other components as well, in order to be fully comprehended in its general context.

Molecular structure can be satisfactorily discussed in terms of the intramolecular distance between two atoms, A and B, and of the various structural parameters controlling this distance. Four cases must be considered, the first three of which are shown in structure I: (1) A and B are adjacent (one bond away), (2) geminal (two bonds away), (3) vicinal (three bonds away), or (4) still more distant. When A and B are adjacent, the only parameter involved is the length of the bond A-B (a one-dimensional parameter). When A and B are geminal, the bond angle A-X-B (a two-dimensional parameter, designated α) is a relevant parameter together with the two bond lengths AX and XB. In the case of A and B being vicinal, the parameters are the bond lengths AX, XY, YB; the bond angles A-X-Y and X-Y-B; and the dihedral angle A-X-Y-B (a three-dimensional parameter, designated Φ). The latter is the angle between the two planes containing atoms AXY and XYB, and intercepting along the X-Y axis (II).

Although presently not as frequent, the case of still more distant atoms has been differentiated from the three other cases (I). Indeed, atoms A and B may be far away along the bonding chain, but close in space (transannular effects in large rings), and nonbonded interactions (see Section 3.3) do occur between them. Nevertheless, the distance between

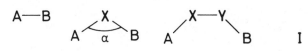

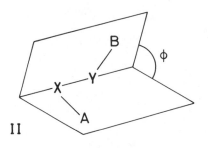

II

A and B is clearly defined by a combination of all bond lengths, bond angles, and dihedral angles encountered along the bonding chain.

Bond lengths, bond angles, and dihedral angles are considered individually in this study in a generally empirical manner, and simple quantum mechanical aspects are discussed only to interpret some of what is found. For further information, the interested reader is referred to the Selected Readings at the end of the book.

3.1 Bond Lengths

When two isolated atomic particles approach each other, the energy of the system undergoes predictable variations. When a bonding molecular orbital can be built by a combination of two atomic orbitals (CAO), the electronic energy (E) versus internuclear distance (R_{AB}) has the general shape displayed in Figure 1. Long-distance attractive interactions and short-distance repulsive interactions equilibrate at an energy minimum corresponding to a well-defined internuclear distance (or bond length) r.

The energy of the attractive interactions is, among others, a function of the atomic orbitals combining to generate molecular orbitals. The formation of a bonding molecular orbital (nonbonding and antibonding orbitals are not discussed) requires maximum overlap and correct symmetry of the atomic orbitals. Generally, rather weak bonds are formed between two s-type atomic orbitals (III). The σ bonds formed along the internuclear axis between one s orbital and one p orbital, and between two p orbitals (IV) are generally stronger, for the simple reason that p orbitals spread their electron density in two general directions only, rather than all around

s   s

III

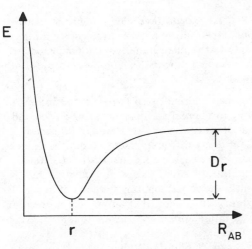

FIGURE 1 Variation of electronic energy E with internuclear distance R_{AB} for a diatomic molecule. The minimal energy at $R_{AB} = r$ corresponds to the formation of a stable molecule. The binding or dissociation energy is obtained from D_r.

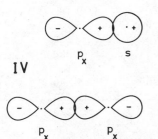

the sphere as an s orbital does. Hybrid atomic orbitals (see Section 3.2) generate even stronger bonds for the same reason. The π bonds resulting from the combination of two p orbitals perpendicular to the internuclear axis (V) are, as a rule, weak when compared to other covalent bonds.

The structural consequences of these bonds' formation are twofold. First, the bond length will depend on the atomic orbitals involved as well

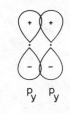

as on the two atoms connected. Second, bonds involving p orbitals (and sp-type hybrid orbitals, to be considered in the next section) show a high degree of directional selectivity due to the maximum overlap condition. (Consequences concerning bond angles and dihedral angles are considered in Sections 3.2 and 3.3, respectively.)

A compilation of bond lengths and bond energies is presented in Table 1. It is important, however, to remember that the values reported are average values of the simplest derivatives and that the actual structural parameters in a given molecule may differ markedly from the values in Table 1. X-ray crystallography (see Section 12.6, and Selected Readings) and quantum mechanical calculations (Ref. 1; see also Section 12.7 and Selected Readings) are currently the most popular methods of determining bond lengths and bond angles. They show that steric and molecular orbital factors play a major role in influencing bond parameters.

TABLE 1 Properties of Organic Bonds (Indicative Values)[a]

Bond	Bond length (Å)	Bond energies	
		in kJ/mol	in kcal/mol
C–H	1.09	420	100
C–C	1.54	350	85
C=C	1.34	670	160
C≡C	1.20	960	230
C–Si	1.87	—	—
C–N	1.47	330	80
C=N	1.30	630	150
C≡N	1.15	920	220
C–P	1.84	—	—
C–O	1.43	380	90
C=O	1.23	730	175
C–S	1.82	250	60
C=S	1.70	540	130
C–F	1.35	450	108
C–Cl	1.78	350	83
C–Br	1.93	290	70
C–I	2.14	230	55
N–H	1.01	420	100
N–N	1.45	290	70
O–H	0.96	460	110
O–O	1.47	210	50

[a] These values are a compilation from the Selected Readings at the end of this text.

Up to now, the chemical bond has been considered only in terms of spatial dimensions. In fact, when adding the temporal dimension, the bond appears to vibrate; it behaves like a harmonic oscillator, just like two masses connected by a spring. This vibration is governed by Hooke's quadratic law. The energy V_r required to distort the bond length is proportional to the square of the linear displacement $(r - r_0)$ away from the equilibrium position r_0:

$$V_r = \frac{k_r}{2} \cdot (r - r_0)^2 \tag{1}$$

where k_r is the bond-stretching force constant. For single C–H and C–C bonds, Eq. (1) takes the approximate values

$$V_r \sim 1450(r - r_0)^2 \quad kJ/mol/\mathring{A}^2$$

$$\sim 350(r - r_0)^2 \quad kcal/mol/\mathring{A}^2$$

A bond deformation (stretching or compression) of $0.05\ \mathring{A}$ thus requires about 3.7 kJ/mol (0.9 kcal/mol).* The values are about twice as much for double bonds. This means that a bond deformation larger than a few percent of the total bond length is a relatively high energy process; covalent bonds are said to be "stiff."

3.2 Bond Angles

The simplest system suitable for discussion of bond angles is a triatomic molecule. The molecule of water is particularly convenient; the oxygen atom has two electrons in the 2s orbital, two electrons in one 2p orbital, and two electrons in the two other 2p orbitals. In order for them to follow the rule of maximum overlap, the two O–H bonds formed between the 1s orbital of H and a 2p orbital of O should display an angle of 90° (VI). The experimental value is 104.5°, a marked discrepancy with the prediction.

A satisfactory model explaining such discrepancies is that of hybridization (or mixing) of atomic orbitals. Instead of considering pure s and p orbitals, the model hypothesizes atomic orbitals to which both s and p orbitals contribute. Thus, two sp, three sp^2, and four sp^3 orbitals result from the hybridization of the one s orbital with, respectively, one, two, and three p orbitals. As stated above, pure p-type orbitals yield a valency

---

*1 cal = 4.1868 J

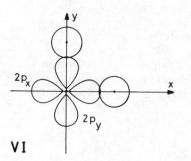

VI

angle of 90° (VI). On the other hand, sp hybridization results in increased angular values. In the case of sp^3, the four hybrid orbitals show regular tetrahedral symmetry (VII) and the valency angle is approximately 109.5° (accurately arc cos - 1/3), close to that found in H_2O. Three sp^2 orbitals lie in the same plane and build valency angles of 120° (VIII). In the case of sp hybridization, the bond angle between the axes of the two sp orbitals is 180°.

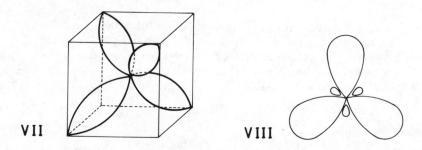

VII   **VIII**

It thus appears that the valency angle increases with increasing s character of the hybrid orbitals (IX). However, it is important to remember that the concept of hybridization is nothing more than a model; while many real cases are satisfactorily approximated, the model has its shortcomings.

IX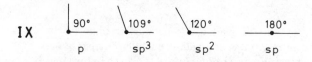

Geminal intramolecular interactions influence bond angles by virtue of steric and electronic effects. Consider, for example, a tetrahedral (sp^3) carbon atom bearing two large (designated L) and two small (designated s) substituents (X). Steric repulsions between the two large substituents will tend to increase the L-C-L angle, with simultaneous decrease of the s-C-s

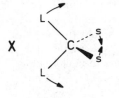

angle; this effect is known as the Thorpe-Ingold effect [1a]. For propane (L = -Me; s = -H), the Me-C-Me angle is thus approximately 112°, while the H-C-H angle is decreased to 107°.

Electronic effects control the valency angle, as discussed above, in that they are related to the state of hybridization of the bonding atomic orbitals. Since s orbitals have their highest electron density at the atomic nucleus and since p orbitals have zero electron density at the nucleus, the p contribution to hybrid orbitals will increase with decreasing electron density at the nucleus. Similarly, an increased electron density at the nucleus increases the s contribution to hybrid orbitals. Two examples will illustrate this point. Substituting an H atom with a chlorine atom in methane yields chloromethane; the electron-withdrawing capacity of the halogen atom will decrease the electron density at the carbon nucleus and increase the p contribution in the bonding sp^3 orbital. As a consequence of this increased p contribution, the H-C-Cl valency angle will decrease to 93°; the s contribution to the three other sp^3 orbitals, of course, increases, and the H-C-H angles increase to 118° (together with a shortening of the C-H bond to 1.05 Å).

The second example is a comparison of H_2S and H_2O. In the former molecule, the electron density on the central atom is smaller than in the latter molecule; the p contribution in the bonding orbitals of H_2S will be larger than in those of H_2O. H_2S has a valency angle of 92.3°, while that of H_2O is 104.5°. (A complementary viewpoint is to consider the increased electrostatic repulsion between the two partially charged H atoms of H_2O as compared to H_2S.)

Up to now, the valency angle has been considered to be the angle between the two internuclear lines. This may not always be so, the classical example being cyclopropane (XI). The internuclear lines build 60° angles, a value markedly smaller than the minimum bond angle of 90°

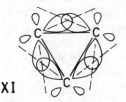

found with pure p orbitals. A convenient way to look at the bonding situation in cyclopropane is to start with pure p orbitals. This would put the C-C bonds as close as is feasible to the internuclear axis, but the overlap would not be optimal (half of the p orbital is on the other side of the atom). Adding some s character to the bonding orbitals increases the bond angle and also the overlap. The C-C bonds are thus bend bonds (XI), also called banana bonds or τ bonds (as opposed to pure σ and π bonds).

The case of cyclopropane raises the question of the energetic aspect of valency angle deformation. Just as bonds vibrate in motions of stretching and compression, so bond angles vibrate in bending motions (radial scissoring motions). This vibration is governed by a Hook's type relationship as are bond length vibrations. The energy V_α required to distort the bond angle is proportional to the square of the angular displacement $(\alpha - \alpha_0)$ away from the equilibrium position α_0:

$$V_\alpha = \frac{k_\alpha}{2} \cdot (\alpha - \alpha_0)^2 \tag{2}$$

where k_α is the bond-bending force constant, and V_α is called the angle strain, or Baeyer strain. For many carbon bond angles, deformations up to approximately 20° obey the empirical form of Eq. (2).

$$V_\alpha \sim 0.04 \, (\alpha - \alpha_0)^2 \quad kJ/mol/deg^2$$

$$\sim 0.01 \, (\alpha - \alpha_0)^2 \quad kcal/mol/deg^2$$

It can be seen that a 1° deformation of a C-C-C angle necessitates no more than 0.04 kJ/mol, while the energy required to deform a bond length by 0.05 Å (about 4 kJ/mol) may induce a 10° bending. Marked bond angle bending will thus be observed quite often as a result of the leveling of various strains within a molecule. See Ref. 2 for the results of a theoretical study in which the energy components associated with distorted geometries at carbon have been calculated using quantum mechanics.

3.3 Dihedral Angles

Variations of dihedral angles Φ (II) have a more profound influence on molecular geometry than variations affecting bond lengths and bond angles. Indeed, it is evident upon consideration of structure II that the vector $\overline{AB}$ (joining atoms A and B) will experience considerable variation in both length and direction with the variation of dihedral angle Φ. The dihedral angle is an important stereochemical parameter, much more so, in fact, than bond length and bond angle which are structural parameters (see beginning of this chapter). These reasons justify the importance given in the following pages to factors controlling dihedral angles and, therefore, torsional isomerism (see Chapters 8 and 9).

The factors to be discussed can be conveniently classified into <u>through-bond interactions</u> involving atoms A-X-Y-B (II), and <u>through-space (non-bonded) interactions</u> involving atoms A and B.

A major factor influencing the dihedral angle A-X-Y-B is the nature of the chemical bond linking X and Y, a factor belonging, in a broad sense, to <u>bonded interactions</u>. A <u>single bond</u> linking X and Y is a σ bond (III, IV); such a bond has no influence per se on the dihedral angle, and the rotation about this bond is said to be <u>free</u> or <u>unrestricted</u>. Restricting factors, as we shall see later, arise from atoms A and B, or other centers further removed.

When X and Y are linked by a <u>double bond</u> (XII), again the σ component does not influence the dihedral angle, whereas the π component of the double bond is a strongly restricting factor. Indeed, maximum overlap of the two p_y atomic orbitals to form the π bond requires that the two axes y and y' be parallel. That this electronic requirement influences the dihedral is evident in the well-known example of ethylene (XIII): the <u>restricted rotation</u> about the C-C double bond blocks the four H in the xz plane. The rotation about the C-C double bond, as measured by the dihedral angle formed by the xy and xy' planes, is a high-energy process (see Section 8.1). Just how high the rotation energy is depends on the two bonded atoms (C=C, C=N, etc.) and on their electronic environment, as assessed by the energies of the molecular orbitals. The stereochemistry of restricted rotation, as well as the ranges of energy levels involved, are considered in more detail in Chapter 8. (See Selected Readings for quantum chemical sources that indicate how such values can be calculated.)

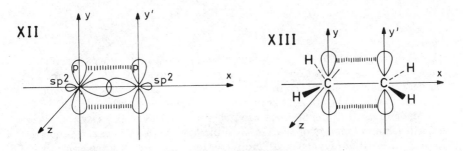

The dichotomy of <u>single bond</u>-<u>free rotation</u> versus <u>double bond</u>-<u>restricted rotation</u> is an example of a two-valued ("either-or") logic, and as such, is unsatisfactory. Intermediate cases have to be accounted for, namely <u>bonds with partial π character</u>.

Such is the case for single bonds adjacent to one (or two) double bond(s), as illustrated by 1,3-butadiene (XIV). Four π electrons exist in this molecule, while quantum mechanics shows the existence of four π-molecular orbitals, two of which (Ψ_1 and Ψ_2) are populated in the ground state (XV). π-Electron delocalization over the 2,3-bond is apparent from Ψ_1; this bond has partial π character, resulting in a shortened sp^2-sp^2 distance (1.46 Å), as compared to the single bond value of 1.54 Å (sp^3-sp^3). Maximum delocalization of the π electrons requires the four carbon atoms to be coplanar, resulting in a slightly restricted rotation about the 2,3-bond (some kilojoules per mole; see Chapter 9, Tables 2, 4, and 6 for typical barrier values).

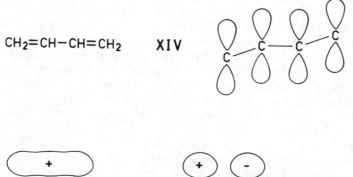

$$CH_2=CH-CH=CH_2 \qquad XIV$$

$$\Psi_1 \qquad\qquad \Psi_2$$

$$XV$$

The extreme case of π-electron delocalization over alternating single and double bonds is that of <u>aromatic rings</u>. The considerable delocalization energy has its corollary in the coplanarity of the ring atoms. Such rings are far from being rigidly planar, as sometimes believed, and minor out-of-plane deformations require little energy. For example, a 5° deviation from planarity of the 1-2-3-4 dihedral angle of benzene requires less than 4 kJ/mol (about 1 kcal/mol), while a 10° deviation requires less than 12 kJ/mol (about 3 kcal/mol) [3].

Partial π-bond character may also arise from a <u>participation of electron lone pairs</u>, particularly nitrogen lone pairs, in bonding to nearby electron sinks. This phenomenon also affects a given dihedral angle.

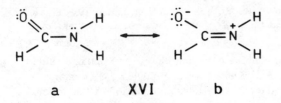

a   XVI   b

Consider, for example, formamide (XVI) depicted as a resonance hybrid. Increased delocalization as indicated in XVI is energetically favorable, while the contribution of XVIb to the hybrid increases as the molecule approaches planarity. Thus, participation of electron lone pairs contributes to the increase of dihedral restriction [4] (see Section 9.4).

Another subtle and yet not fully understood through-bond interaction is hyperconjugation, which occurs in systems such as $CH_3-C=C$. The interaction of the electron densities of the methyl C-H bonds with the adjacent π-electron densities constitutes hyperconjugation, the methyl group acting as an electron density donator. Donation increases as the H-C=C dihedral angle reaches energetically unfavorable values, showing hyperconjugation to be an ingredient in the barrier of rotation [5] (see Section 9.4). A clear and illuminating discussion on conjugative interactions can be found in the work of Epiotis and collaborators [5a].

The through-bond interactions discussed above are either high- or low-energy interactions, requiring, respectively, many tens of kilojoules per mole (e.g., carbon-carbon double bonds, Section 8.1), and a few kilojoules per mole (e.g., single bonds with partial π character, Section 9.4), in order to overcome the rotation barrier. The nonbonded interactions to be discussed now are mostly low-energy processes involving a few kilojoules per mole; as such, they will be of significance in controlling the dihedral angle A-X-Y-B only when the through-bond interactions are of comparable energy.

Through-space interactions can be divided into a nuclear repulsion component, and stabilizing or destabilizing interactions of electronic nature. The well-known electrostatic interactions provide a simple and empirical way of describing some electronic interactions. Classically, the energy of electrostatic interactions, also called Coulomb interactions, is described by functions, including, among others, the following:

The two interacting centers (more precisely their electric charge, q, their dipole moment, μ, or their polarizability, a)

The reciprocal of the dielectric constant, D, of the medium separating the two centers (i.e., D^{-1})

The reciprocal of the distance, r (at the nth power) between the two centers (i.e., r^{-n})

When the power of the distance r is -1, the interaction energy describes on the ionic bond. When r is at the power -2, -4, or -6, one speaks of higher-order interactions.

Ionic interactions between centers carrying charges of opposite signs display an energy curve similar to that shown in Figure 1 for covalent bonds. The attractive part of the curve follows an equation of the general shape

$$E = \frac{q_A \cdot q_B}{D \cdot r} \tag{3}$$

as discussed above. It should be remembered, to avoid confusion, that when a force rather than an energy is considered, r is at the -(n + 1)th power. When the distance r becomes smaller than a threshold value known as the sum of the van der Waals radii of the two atoms, the nuclear repulsion starts to predominate and increases esponentially, as in Figure 1. (Steric repulsions are considered at the end of the chapter.)

Attractive ionic interactions may arise intramolecularly between groups such as $-CO_2^-$ and $-N^+R_3$. Interactions repulsive for all values of r will arise between centers having charges of same sign, e.g., two positively charged nitrogen atoms. From Eq. (3), it is apparent that ionic interactions are long-range forces which are minimized in highly polar solvents such as water.

To the Coulomb interactions of higher order belong the ion-dipole and the van der Waals interactions, which are summarized in Table 2.

TABLE 2 Higher Order Electrostatic Interactions

Name of interaction	Interacting centers	Distance term in energy function
Ion-dipole	Ion - permanent dipole	r^{-2}
	Ion - induced dipole	r^{-4}
van der Waals forces		
Keesom forces (orientation forces)	Permanent dipole - permanent dipole	r^{-6}
Debye forces (induction forces)	Permanent dipole - induced dipole	r^{-6}
London forces (dispersion forces)	Instantaneous dipole - induced dipole	r^{-6}

These forces are intense, but of short range, since their intensity decreases extremely rapidly with increasing distance. Since dipoles are oriented by nature, interactions involving them will vary with orientation. For example, an ion-dipole interaction will be maximal when one end of the dipole points directly toward an ionic center of opposite sign, and will decrease as the dipole axis deviates from the internuclear axis. This directional aspect may have stereochemical consequences.

An interesting example of ion-dipole interactions has been reported for vicinal -O- (ether group) and $-N^+R_3$ centers; the preferred O-C-C-N dihedral angle was shown to be one allowing the closest permissible approach between the interacting centers [6]. Also, well-known attractive interactions occur between halogen atoms (as part of a dipole carrying an electron excess) and hydrogen atoms. Recently, nonbonded, attractive interactions have been characterized between aliphatic and aromatic systems [6a]. The detected C-H$\cdots\pi$ interactions are quite weak (about 4 kJ/mol at the most) but they may play a marked role in many molecular systems.

Repulsive electrostatic interactions have also been documented. For example, a strong dipole-dipole repulsion exists between a carbonyl oxygen and a halogen atom (e.g., fluoroacetone, XVII) [7]. Destabilizing interactions also exist between the electron lone pair of a given substituent, and the π electrons of a C-C double bond. As expected, the interaction is decreased in polar solvents and increased in apolar solvents [8,9].

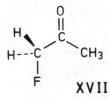

XVII

Of major significance in chemistry and biochemistry is the hydrogen bond (H bond) [e.g., 10,11]. This attractive interaction occurs between a proton donor group (A-H) and an electron donor (B) group, and may be either intermolecular or intramolecular. Proton donors are groups containing an electron-poor hydrogen atom, while electron donors are groups containing either lone pairs on a heteroatom or a π-electron system.

Increasing the electronegativity of the donor atom A increases the strength of the H bond (i.e., S<N<O). With a given A-H donor, the strength of H bonds increases in the order of acceptors B: π electrons < alkyl halides < nitriles < thiocarbonyls and thioethers < carbonyls and ethers < amines. Representative H bonds are listed in Table 3 together

TABLE 3 Representative Hydrogen Bonds (A-H$\cdots$B)
  (Average A-B Distances; Negative Enthalpies)

Strong H bonds (2.4-2.9 Å; 16-35 kJ/mol)

$-O-H\cdots O\langle$ (about 2.7 Å)

$-O-H\cdots S\langle$

$-O-H\cdots N\langle$ (about 2.8 Å)

$\rangle N-H\cdots O\langle$ (about 2.9 Å)

Weaker bonds (3.0-3.5 Å; 8-16 kJ/mol)

$\rangle N-H\cdots N\langle$ (about 3.1 Å)

$\rangle N-H\cdots S\langle$ (about 3.4 Å)

$-O-H\cdots F-$

Weak bonds

Cl-C-H$\cdots$O (with Cl substituents) $-O-H\cdots$ C=C

$\rangle N-H\cdots \pi$ $-O-H\cdots$ C≡C

---

with some A-B distances and enthalpy terms; however, these data have
only indicative value, since H bonds are strongly influenced by molecular
(electronic and steric), and environmental (solvent, temperature) factors.

The hydrogen bond is a complex one involving interactions of several
types, i.e., dipole-dipole interactions and other Coulombic interactions,
charge transfer (delocalization), van der Waals forces, and repulsive
forces and their coupling. Among the stabilizing components, the electrostatic terms predominate. A recent theoretical energy decomposition
study allows much insight into the origin of hydrogen bonding and points to
this bond's unique character [12].

That repulsive interactions are components of the H bond can be
understood quite simply from the fact that the A-B distance in an A-H$\cdots$B

bond is usually much smaller than the sum of the van der Waals radii $r_A + r_B + 2r_H$. In fact, quite often this A-B distance is so small that the H atom is almost nonexistent in terms of van der Waals radii, i.e., resembling a free proton. Indeed, the electron density around the H atom is very small and, therefore, the van der Waals radius is minute when compared, for example, to a hydrogen in a hydrocarbon.

Some structural factors influencing the strength of a hydrogen bond are interatomic distances, and directionality and linearity [12], as also encountered in dipole-dipole interactions. A colinear configuration of A, H, and B is expected to have the lowest potential energy. However, due to the small energies involved, large deviations from colinearity will correspond to only a very few kilojoules per mole or less.

Intramolecular H bonds result in the formation of cyclic structures, whereas intermolecular H bonds form dimers (or oligomers). Up to now, hydrogen bonds have been considered globally, but it is obvious that only intramolecular H bonds are relevant to the present chapter. Several differences exist between the two forms. In thermodynamic terms, enthalpy values for intramolecular H bonds are smaller than for the intermolecular H bonds, because the intramolecular constraints prevent the atoms from assuming optimal distances and angles. On the other hand, the entropy factor is less favorable in intermolecular bonds because the independent translational motions of the two molecules have to be "frozen out" in order to create the bond, a situation not encountered in intramolecular bonds. As a result, the net free energy of bonding is often larger for intramolecular H atoms. Another important difference lies in the behavior toward dilution; decreasing concentration decreases the fraction of intermolecular H bonds relative to unbonded structures while intramolecular H bonds are, in principle, insensitive to dilution.

In general terms, intramolecular hydrogen bonds, whenever feasible, will play a marked role in controlling the three-dimensional structure of molecules.

The description of electronic interactions in terms of electrostatic potentials is a rather crude empirical approach allowing coarse predictions, if any, and gives no insight into the relevant phenomena. Quantum chemical models have been and are being developed to treat nonbonded interactions; they consider, among others, one-electron attractive interactions and two-electron repulsive potentials [13-15]. In particular, the one-electron molecular orbital (OEMO) theory appears as remarkably promising and fruitful in quantitatively describing and predicting attractive interactions (see Ref. 16).

But, theories such as the OEMO model do not explicitly treat interelectronic repulsions [16]. Internuclear repulsive potentials are also a significant component of nonbonded interactions [17]. Interelectronic and internuclear repulsions are called steric effects, or steric repulsions. They are related to the concept of the van der Waals radius of atoms. When two atoms, either in the same or in different molecules, approach each other, the potential energy curve shows a minimum. At this minimum, the internuclei distance is considered by most theoreticians and chemists to be the sum of the two van der Waals radii [18]. Beyond the threshold of the van der Waals radii, the potential energy increases exponentially with decreasing distance, and a "hardness parameter" is defined [18]. Therefore, intuitively, it is more appropriate to consider atoms not as hard spheres, but as rubber balls showing an "elastic" behavior. Further, for atoms with few electrons, and especially for hydrogen, the van der Waals radius varies with the direction of approach, suggesting a somewhat ellipsoidal shape. Despite these limitations, van der Waals radii are of the greatest significance when considering molecular structures; some common values are listed in Table 4.

For crystallographers, the van der Waals radius relates to the distance of closest approach in crystals. Due to intermolecular packing forces, the definition implies radii which are smaller (by approximately 0.3 Å) than van der Waals radii applied to isolated atoms. For hydrogen, for example, the crystallographers' value is 1.2 Å and this value has long been the only one used when assessing the steric effect of hydrogen atoms [18]. The more realistic value of 1.5 Å indicates that interhydrogen steric repulsions may play a greater role in controlling dihedral angles than has been assumed until recently [19,20]. This value of 1.5 Å applies,

TABLE 4 van der Waals Radii in Å[a]

				H 1.50
B 1.80	C 1.85 sp, sp^2 1.75 sp^3	N 1.70	O 1.65	F 1.60
	Si 2.10	P 2.05	S 2.00	Cl 1.95
	Ge 2.25	As 2.20	Se 2.15	Br 2.10
	Sn 2.40			I 2.25

[a]From Allinger [18].

however, to hydrocarbon hydrogens, which are bigger and harder than hydrogens adjacent to electron-withdrawing atoms such as oxygen and nitrogen (see, for example, Ref. 21).

The phenomenon of steric interactions (usually called steric effects) can also be investigated by means of quantitative structure-reactivity relationships. This approach has already proven useful in assisting the influence of steric effects on barriers to internal rotation [22], and may lead to a new conception of intramolecular steric interactions.

The energy variation implicated in the torsion of a given dihedral angle Φ is the sum of all involved interactions, through-bond and through-space. In its simplest form, the torsional energy may be written as

$$V_{\Phi} = 1/2\, V_0\; (1 + \cos[n\Delta\Phi]) \tag{4}$$

where V_0 is the torsional energy barrier, n is the periodicity, and V_{Φ} is the torsional strain [23]. These aspects are given due consideration in the chapters dealing with torsional isomerism (Chapters 8 and 9).

References

1. H. M. Niemeyer, Tetrahedron, 33:1369 (1977).

1a. E. L. Eliel, N. L. Allinger, S. J. Angyal, and G. A. Morrison, Conformational Analysis, Wiley, New York, 1965, p. 191.

2. K. B. Wiberg and G. B. Ellison, Tetrahedron, 30:1573 (1974).

3. H. Wynberg, W. C. Nieuwpoort, and H. T. Jonkman, Tetrahedron Lett., 4623 (1973).

4. D. A. Kleier, J. H. Hall, Jr., T. A. Halgren, and W. N. Lipscomb, Proc. Natl. Acad. Sci. USA, 71:2265 (1974).

5. A. Liberles, B. O'Leary, J. E. Eilers, and D. R. Whitman, J. Am. Chem. Soc., 94:6894 (1972).

5a. N. D. Epiotis, W. R. Cherry, S. Shaik, R. L. Yates, and F. Bernardi, Structural Theory of Organic Chemistry, Springer-Verlag, Berlin, 1977, part IV.

6. E. L. Eliel and F. Alcudia, J. Am. Chem. Soc., 96:1939 (1974).

6a. Y. Kodama, K. Nishihata, M. Nishio, and N. Nakagawa, Tetrahedron Lett., 2105 (1977).

7. E. B. Wilson, Chem. Soc. Rev., 1:293 (1972).

8. J. B. Lambert, R. R. Clikeman, and E. S. Magyar, J. Am. Chem. Soc., 96:2265 (1974).

9. J. B. Lambert and R. R. Clikeman, J. Am. Chem. Soc., 98:4203 (1976).

10. S. N. Vinogradov and R. H. Linnell, Hydrogen Bonding, Van Nostrand-Reinhold, New York, 1971.

11. L. C. Allen, A simple model of hydrogen bonding, J. Am. Chem. Soc., 97:6921-6940 (1975).

12. H. Umeyama and K. Morokuma, J. Am. Chem. Soc., 99:1316 (1977).

13. N. D. Epiotis, D. Bjorkquist, L. Bjorkquist, and S. Sarkanen, J. Am. Chem. Soc., 95:7558 (1973).

14. N. D. Epiotis, R. L. Yates, and F. Bernardi, J. Am. Chem. Soc., 97:5961 (1975).

15. F. A. Van-Catledge, J. Am. Chem. Soc., 96:5693 (1974).

16. N. D. Epiotis, W. R. Cherry, S. Shaik, R. L. Yates, and F. Bernardi, Structural Theory of Organic Chemistry, Springer-Verlag, Berlin, 1977, parts I and II.

17. I. G. Csizmadia, G. Theodorakopoulos, H. B. Schlegel, M. H. Whangbo, and S. Wolfe, Can. J. Chem., 55:986 (1977).

18. N. L. Allinger, Calculation of molecular structure and energy by force-field methods, in Advances in Physical Organic Chemistry, vol. 13 (edited by V. Gold and D. Bethell), Academic Press, London, 1976, pp. 1-83.

19. D. H. Wertz and N. L. Allinger, Tetrahedron, 30:1579 (1974).

20. S. Fitzwater and L. S. Bartell, J. Am. Chem. Soc., 98:5107 (1976).

21. N. L. Allinger and D. Y. Chung, J. Am. Chem. Soc., 98:6798 (1976).

22. M. Charton and B. Charton, J. Am. Chem. Soc., 97:6472 (1975).

23. K. Mislow, Introduction to Stereochemistry, Benjamin, New York, 1966, pp. 35-36.

4

The Classification of Isomeric Structures

4.1 Classification of Stereoisomers (Whole Molecules)

Two isomers can be defined as distinct molecular entities sharing a common characteristic in their chemical structure. Thus, structural isomers are distinct chemical species which have the same molecular formula.

A single molecular formula may, however, include several chemical constitutions (defined as the sequential arrangement of atoms regardless of direction in space). Structural isomers differing in constitution (for example, 1-propanol and 2-propanol) are called constitutional isomers. On the other hand, structural isomers having identical constitutions differ in the spatial arrangement of their atoms and are called stereoisomers (Scheme 1).

A precise classification of constitutional isomers remains to be undertaken. Broad and overlapping subgroups include positional isomers (regioisomers) and tautomers (H tautomers, ring-chain tautomers, etc.).

But the classification of stereoisomers is well-defined, and is usually done according to two distinct and independent criteria, namely, symmetry and energy criteria.

Symmetry classification divides stereoisomers into enantiomers and diastereoisomers (Scheme 1). Either two stereoisomers are related to each other as object and nonsuperimposable mirror image, or they are not. In the former case, the two stereoisomers share an enantiomeric relationship and are called enantiomers. This implies that the molecules be dissymmetric (chiral) (see Section 1.2); chirality is the necessary and sufficient condition for the existence of enantiomers. For example, the two isomeric forms, structures I and II, as drawn, are nonsuperimposable mirror images, i.e., enantiomers.

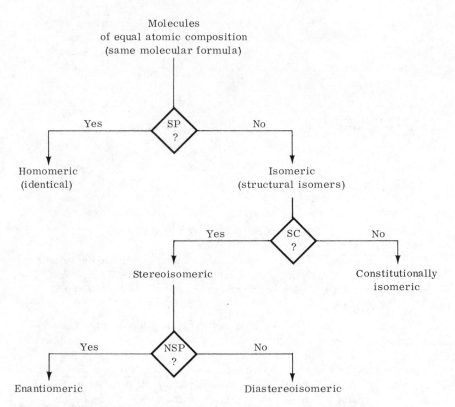

SCHEME 1 The conventional classification of relations between isomeric molecules. SP = superimposable; SC = same constitution; NSP = non-superimposable mirror images. Adapted from K. Mislow [10].

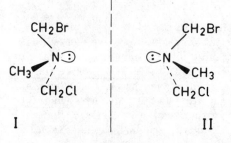

I

II

Since enantiomers are optically active and show optical rotations of identical amplitude, but opposed signs (see Section 12.3), they are also called <u>optical isomers</u>.* The prefixes (+)- and (-)- for dextrorotatory and levorotatory, respectively, allow differentiation of configurationally stable enantiomers whose absolute configuration may not be known. But, the prefixes d and l should never be used because of possible confusion with the D and L nomenclature; see Section 5.3.

Equimolar parts of two enantiomers compose a particular mixture called a <u>racemic modification</u> (see Section 12.5), or a (±)-pair. When the name of a chiral compound is given without a prefix indicative of either absolute configuration or optical rotation, a racemic form is assumed. The prefix (±)- should, however, be used in ambiguous cases.

Stereoisomers which are not enantiomers are <u>diastereoisomers</u> and are said to share a <u>diastereoisomeric relationship</u>. Such is the case for cis- and <u>trans</u>-1,2-dichloroethylene (III and IV), two planar achiral stereoisomeric molecules. While a molecule can have one and only one enantiomer, it may have several diastereoisomers. However, two stereoisomers cannot, at the same time, be enantiomers and diastereoisomers of each other. In other words, and this is a major stereochemical rule, the enantiomeric and diastereoisomeric relationships are mutually exclusive. (Several additional examples will be discussed in the following chapters.)

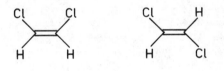

III   IV

The atoms within a chiral molecule have exactly the same relative positions and interactions as they have within the enantiomeric molecule, as can be seen when considering structures I and II. The energy contents of two enantiomers are thus identical, and they will display identical chemical and physical properties. Diastereoisomers, though, have different relative positions of atoms, as exemplified by structures III and IV, which show different Cl/Cl and H/H distances. This means diastereoisomers differ in energy content and in every physical and chemical property, however minute the difference may be in some cases.

These fundamental differences between enantiomers and diastereoisomers are of the utmost significance and will be given due attention in Chapter 12, and especially in Section 12.1.

---

*A few dissymmetric molecules do not exhibit detectable optical activity; they are de facto optically inactive (see Section 12.3). The popular term "optical isomer" is, therefore, not totally appropriate.

Energy classification is concerned with the energy necessary to convert a given stereoisomer into its isomeric form; here, the energy barrier separating two stereoisomers becomes the criterion of classification. In quantitative terms, stereoisomers separated by a "high"-energy barrier are called configurational isomers, while a "low"-energy barrier separates conformational isomers (conformers).

Within the general energy criterion, a precise differentiation between configuration and conformation is rendered difficult by the variety of existing definitions.*

In the author's opinion, it is gainful to apply the energy criterion consistently to all stereoisomers, whatever the intramolecular process resulting in their interconversion (either bond rotation, inversion, or pseudorotation). Reasoning with specific cases, the enantiomers I and II are interconvertible by a typical low-energy inversion process requiring several kilojoules per mole [e.g., 2]; they are considered as conformers, and, more precisely, as conformational enantiomers. Compounds III and IV, on the other hand, can be conceptually interconverted by a torsion involving π-bond breaking, a typical high-energy process (see Section 8.1). They are configurational isomers, or, in better terms, configurational diastereoisomers.

When applying the conformation-configuration dichotomy to stereoisomers, the main difficulty arises from the fact that there is no well-defined limit separating the two concepts. If conformation refers to a "low"-energy barrier, and configuration to a "high"-energy barrier, how can intermediate cases be classified? Should an arbitrary cutoff point be set in a continuous series of values, calling for accurate knowledge of energy barriers as a condition of classification?

---

*The International Union of Pure and Applied Chemistry (IUPAC) aptly summarized these definitions, concluding that "the time does not seem ripe to legislate . . ."[1]. The original agreement considers as conformers those stereoisomers interconverted by rotations about single bonds, while interconversion of configurational isomers requires bond breaking; the energy of activation is usually low in the first case and high in the second case. An extension of this usage also considers conformational isomerism to involve rotation about bonds of fractional order between one and two. But, then, does torsional isomerism about any genuine double bond resort to conformation (which would be absurd in the author's mind), or to configuration? And where should the limit be set in terms of bond order? Further, all stereoisomers interconverted by an inversion process (invertomers) tend, at times, to be considered as configurational isomers, independent of the energy barrier.

Rather, the boundary between conformation and configuration should be viewed as a broad energy range encompassing the value of 80 kJ/mol (20 kcal/mol), i.e., the limit of fair stability under ambient conditions.* Such fuzziness of criteria, far from being a drawback, may one day open new vistas and ways of thinking. As an example of this, consider the newly developed "fuzzy sets" theory in the field of mathematics [5-8].

The above classification of stereoisomers, according to the two independent criteria of symmetry and energy, will be used throughout this book and is summarized in Figure 1. Placing all cases of stereoisomerism in a "box," a sharp vertical division discriminates between enantiomers and diastereoisomers, while a broad horizontal division separates conformers and configurational isomers, with allowance for some overlap between the two fields.

---

*Separation of isomers at room temperature (298°K) requires mean lifetimes of several hours. This is equivalent to a rate constant for interconversion of $k < 10^{-4}$ sec^{-1}, and corresponds approximately to a free energy of activation of $\Delta G^{\ddagger} > 96$ kJ/mol [3], as calculated by Eyring equation [4]

$$k = K(k_B T/h)e^{-\Delta G^{\ddagger}/RT} \tag{1}$$

where K is the transmission coefficient, k_B is the Botzmann constant, h is the Planck constant, R is the gas constant, and T is the temperature in degrees Kelvin.

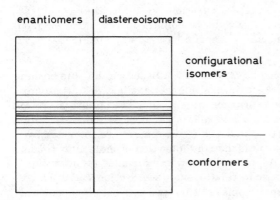

FIGURE 1 Summary of the classification of stereoisomers.

4.2 Classification of Steric Relationships
Between Molecular Fragments

Molecular fragments may show steric relationships as do whole molecules [9]. When such fragments are considered in isolation, i.e., separated from the remainder of the molecule, morphic relationships exist. When the partial structures are considered in an intact molecule (or in different intact molecules), one speaks of topic relationships.

Morphic and topic analyses are based on criteria which are analogous to those used in isomer classification (Section 4.1). This is evidenced by Scheme 2 and its close similarity with Scheme 1. Topic relationships are of particular significance in stereochemistry and will be considered again at some length in the context of prostereoisomerism (Chapter 11).

4.3 The Isometry-Based Classification of
Isomeric Structures

The conventional classification presented in Scheme 1 is based upon the bonding connectivity of atoms. As a result, a critical discrimination is made between isomers which are identically connected (stereoisomers) and those which are not (constitutional isomers).

This categorization, although easily understood and widely accepted, is not fully satisfactory, as aptly demonstrated by Mislow [10]. Indeed, we have seen above and shall discuss at length later (Section 12.1) the fact that enantiomers have identical chemical and physical properties, while diastereoisomers differ in every physical and chemical property. As such, the latter resemble constitutional isomers, with which they have more in common than with enantiomers. Furthermore, the conventional classification separates homomers and enantiomers, a peculiar result if one considers that enantiomers behave as homomers in an achiral environment, i.e., when no chiral means of discrimination is applied (Section 12.1).

The classification proposed by Mislow is not based on the sole bonding connectivity of atoms, but on the pairwise interactions between all atoms in a molecule. The usual way of representing a complete molecular set of interactions takes the form of a matrix, and in our context the most convenient parameters to use to fill the boxes in the matrix are interatomic distances. Let us consider some isomeric molecules of molecular formula C_4H_5Cl. The allenes V and VI are asymmetric molecules (point group C_1) and they share an enantiomeric relationship (see Section 7.1). These two molecules are constitutional isomers of the two 1,3-butadiene derivatives, structures VII and VIII. The latter are achiral molecules (group C_s) and share a diastereoisomeric relationship. Thus, V is an enantiomer of

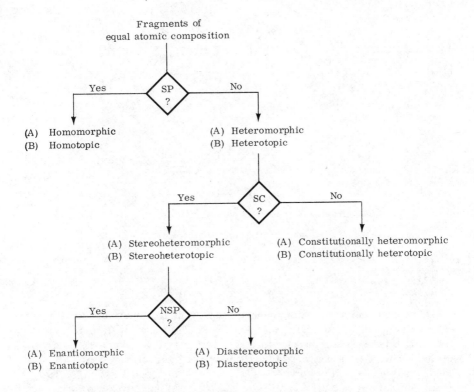

SCHEME 2 The conventional classification of relations between morphic or topic molecular fragments. (A) Morphic relations: fragments isolated from the rest of the molecule; (B) Topic relations: fragments in an intact molecule, or in different intact molecules; SP = superimposable; SC = same constitution; NSP = nonsuperimposable mirror images. Adapted from K. R. Hanson [9] and K. Mislow [10].

VI and a constitutional isomer of VII and VIII, while VII is a diastereoisomer of VIII and a constitutional isomer of V and VI.

The distance matrices of compounds V to VIII are given in Table 1 (assuming the planar conformations shown for VII and VIII). In order to simplify the table, only heavy atoms are considered; the hydrogen-suppressed matrices obtained in this manner are simpler and easier to grasp than the complete matrices, without loss of exactness.

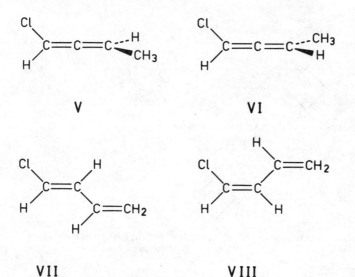

V   V I

VII   VIII

If we consider the bonding connectivity in the four matrices (under-
scored data), we see that it is identical for V and VI, as well as for VII
and VIII; this confirms the fact that V and VI are stereoisomers, as are
VII and VIII. Let us now consider all pairwise interactions, i.e., all
bonded and nonbonded distances. The matrices of V and VI are identical,
while the matrices of VII and VIII are not (Cl-C(3) and Cl-C(4) differ).
This operation of comparison is called isometry and is defined as a transformation which preserves the lengths of all line segments between pairs
of points (atoms) [10]. Compound V is isometric with VI, but anisometric
with VII and VIII; compound VII is anisometric with V, VI, and VIII.
More precisely, their molecular graphs are isometric or anisometric,
respectively.

Applying the isometry operation to isomeric molecules, two major
categories, isometric and anisometric molecules, are obtained (Scheme 3).
Isometric molecules are homomeric or enantiomeric depending on whether
or not they are superimposable. Anisometric molecules are diastereoisomeric or constitutionally isomeric depending on whether or not they
have the same constitution.

The isometry-based classification (Scheme 3) is unambiguous and
meets the criticisms addressed to the conventional classification.
Although not shown here, it is also applicable to molecular fragments in
an analogous manner [10]. The conventional terms "structural isomerism"
and "stereoisomerism" have no place in the newly proposed classification.
The future might conceivably see the term "stereoisomer" fall into
obsolescence or change its meaning. But, for the time being

TABLE 1 Hydrogen-Suppressed Distance Matrices of Structural
Isomers C_4H_5Cl [a]

Cl - C(1) - C(2) - C(3) - C(4)

(S)-1-chloro-1,2-butadiene (V)

	Cl	C(1)	C(2)	C(3)	C(4)
Cl	—	1.75	2.7	3.85	4.75
C(1)		—	1.3	2.6	3.65
C(2)			—	1.3	2.45
C(3)				—	1.55
C(4)					—

(R)-1-chloro-1,2-butadiene (VI)

	Cl	C(1)	C(2)	C(3)	C(4)
Cl	—	1.75	2.7	3.85	4.75
C(1)		—	1.3	2.6	3.65
C(2)			—	1.3	2.45
C(3)				—	1.55
C(4)					—

trans-1-chloro-1,3-butadiene (VII) [b]

	Cl	C(1)	C(2)	C(3)	C(4)
Cl	—	1.75	2.65	4.1	5.15
C(1)		—	1.3	2.5	3.65
C(2)			—	1.5	2.5
C(3)				—	1.3
C(4)					—

cis-1-chloro-1,3-butadiene (VIII) [b]

	Cl	C(1)	C(2)	C(3)	C(4)
Cl	—	1.75	2.65	2.95	4.3
C(1)		—	1.3	2.5	3.65
C(2)			—	1.5	2.5
C(3)				—	1.3
C(4)					—

[a] Measured on Dreiding molecular models.

[b] Conformation as shown in the diagram.

"stereoisomerism" is still going strong, and this text remains dedicated
to "stereochemistry"!

The energy classification of stereoisomers discussed in Section 4.1
remains unaffected by the isometry-based classification. The latter
indeed is symmetry-based, just like the classification it should replace
(Scheme 1). When, by analog with Figure 1, the isometry- and energy-
based classifications are combined, two classes of constitutional isomers
appear, namely those separated by low- and high-energy barriers.
Examples of the high-energy category are very common, while H tautomers
(protomers) exemplify the low-energy category.

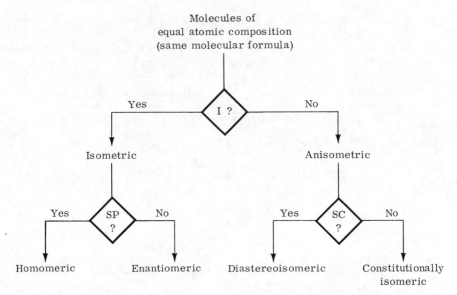

SCHEME 3 The isometry-based classification of relations between isomeric molecules. I = isometric; SC = same constitution; SP = superimposable. Adapted from K. Mislow [10].

References

1. IUPAC Commission on Nomenclature of Organic Chemistry, 1974 recommendations for section E, fundamental stereochemistry, Pure Appl. Chem, 45:11 (1976).

2. C. H. Bushweller, W. G. Anderson, P. E. Stevenson, and J. W. O'Neil, J. Am. Chem. Soc., 97:4338 (1975).

3. H. O. Kalinowski and H. Kessler, Fast isomerizations about double bonds, in Topics in Stereochemistry, Vol. 7 (edited by N. L. Allinger and E. L. Eliel), Wiley, New York, 1973, pp. 295-383.

4. H. Eyring, Chem. Revs., 17:65 (1935).

5. L. A. Zadeh, Fuzzy sets, Information and Control, 8:338 (1965).

6. L. A. Zadeh, Fuzzy sets and systems, Proceedings of the Symposium on System Theory, Polytechn., New York, 1965, pp. 29-39.

7. L. A. Zadeh, Probability measures of fuzzy events, J. Math. Anal. Appl., 10:421 (1968).

8. L. A. Zadeh, in Biological Applications of the Theory of Fuzzy Sets and Systems in Biocybernetics of the Central Nervous System (edited by L. C. Proctor and L. C. Tille), Brown & Co., Boston, 1969, pp. 199-206.

9. K. R. Hanson, Concepts and perspectives in enzyme stereochemistry, Annu. Rev. Biochem., 45:307 (1976).

10. K. Mislow, Bull. Soc. Chim. Belg., 86:595 (1977).

5

Stereoisomerism Resulting from
a Single Center of Chirality

A center of chirality within a molecule consists of any suitable atom substituted in such a way as to be nonsuperimposable on its mirror image. Such an atom is often designated "asymmetric," even though it is the assembly of the atom, plus its ligands, which makes it asymmetric.* In the present chapter, we shall consider molecules containing one such center of chirality.

Chiral centers may be tricoordinate, tetracoordinate, pentacoordinate, or may even possess a higher degree of coordination. Pentacoordinate centers represent a stereochemical field of great complexity whose study is relatively recent, and which will not be discussed here. In order to justify the fascination and hopes of stereochemists, let us simply keep in mind that a center substituted by five different ligands in a trigonal-bipyramidal geometry (I) can conceivably exist as ten diastereoisomeric pairs of enantiomers [1,1a,1b]!

5.1 Chiral Tetracoordinate Centers

A central atom bearing four different substituents lies at the center of a chiral tetrahedral structure and may be termed a chiral tetracoordinate center. Indeed, such an assembly is asymmetric (group C_1) and has one, and one only, stereoisomer which is its enantiomeric form (II). The reversible conversion of structure IIa to IIb may result from two different mechanisms. One of the four substituents may be cleaved from the central atom, to which it then rebinds in a position different from the original. The transition state of such a mechanism involves coplanarity of the central

---

*In the case of molecular asymmetry, where there is no asymmetric atom, one can have a geometrical center of chirality which is not located on the atom.

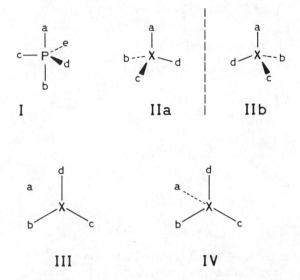

I   IIa   IIb

III   IV

atom and the three remaining ligands (III). Alternatively, no bond cleavage occurs; the central atom goes through the plane of three of its ligands, and the fourth one swings around later or earlier (IV). These two mechanisms, in fact, represent limiting situations between intermediate possibilities which involve a planar transition state and bond elongation(s) (partial bond cleavage).

The reversible inversion process is called <u>racemization</u> when it tends to irreversibly transform one of the isolated enantiomers into the racemic form. This is most often a high-energy process, since all conceivable transition states can be reached only with considerable energy expense. In this case, the enantiomers IIa and IIb are <u>configurational isomers</u> and are stable at room temperature.

To put the above discussion in a more concrete perspective, we must now examine the nature of the central atom X (structure II) and its four different substituents a, b, c, d. Most frequently, X is a <u>carbon atom.</u> The chirality of this center is retained even when two substituents show as little difference as a hydrogen atom and its isotope deuterium. Proof of chirality is found in optical rotation, as exemplified by $(+)$-α-d-ethylbenzene (V) [2]. Optical activity and structural features responsible for this phenomenon are discussed in Section 12.3.

Another element which has significance as a chiral center is the nitrogen atom. The <u>quaternary nitrogen</u> substituted with four different ligands (VI)* is chiral and configurationally stable, as are <u>N-oxides</u> (amine-oxides)

---

*a $\neq$ b $\neq$ c $\neq$ d $\neq$ H for structures VI, VIII, IX, and X.

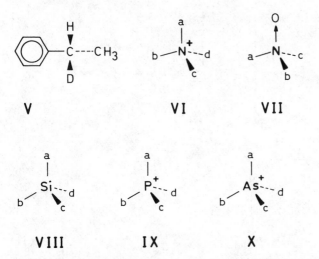

V   VI   VII

VIII   IX   X

of the type depicted (VII). Other tetrahedral centers include <u>silicon</u> (silanes, VIII) and <u>germanium</u> (germanes) derivatives, as well as <u>phosphonium</u> (IX) and <u>arsonium (X)</u> salts. The resolution and configurational stability of such compounds are well-documented.

5.2 Chiral Tricoordinate Centers

An atom bonded to three different ligands possesses a chiral center if either of two conditions is fulfilled [3]. First, the central atom is not coplanar with the three substituents; the resulting <u>pyramidal geometry</u> is chiral (XI). The conversion of structure XIa to XIb occurs via a planar transition state and is called <u>pyramidal inversion</u>, or simply inversion. Second, a planar center may carry a substituent devoid of conical symmetry (XII); however, this system racemizes by rotation about the a-X bond and resorts to torsional isomerism (to be discussed in Chapter 9).

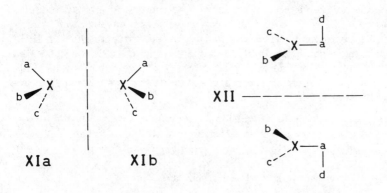

XIa   XIb   XII

TABLE 1 Inversion Barriers for Some Molecules Involving First-
and Second-Row Tricoordinate Central Atoms

Compound	Barrier in kJ/mol (kcal/mol)	
	Experimental[a]	Calculated[b]
CH_3^-	—	22.9 (5.46)
SiH_3^-	—	166.0 (39.6)
NH_3	24.7 (5.9)	21.3 (5.08)
CH_3NH_2	20.1 (4.8)	—
$(CH_3)_2NH$	18.4 (4.4)	36.0 (8.6)
$(CH_3)_3N$	34.4 (8.2)	—
PH_3	—	156.0 (37.2)
OH_3^+	7.1 (1.7)	—
SH_3^+	—	126.0 (30.0)

[a]Taken from a compilation given in Ref. 3a.

[b]Ab initio calculations, taken from a compilation given in Ref. 3b.

While the tetracoordinate centers were found to be configurationally
stable, considerable variation exists among tricoordinate centers in their
rates of inversion. Stable trisubstituted centers include sulfonium salts
(XIII) and sulfoxides (XIV), phosphines (XV), and arsines and stibines.
However, tricoordinate derivatives of carbon, oxygen, and nitrogen (first-
row atoms) experience fast inversion and are configurationally unstable;
they are, therefore, to be viewed as conformationally chiral. Table 1
lists the inversion barriers of some tricoordinate first- and second-row
atoms; although the reported compounds are achiral, they are nevertheless
interesting as models. Carbanions (XVI) racemize readily (e.g., the
methyl anion, Table 1) while exceptions such as the cyclopropyl anion are
known (barrier of inversion about 88 kJ/mol) [3]. Carbon radicals (XVII)
and carbonium ions (XVIII) are usually close to planarity and tend to be
achiral independent of their substituents. Oxonium salts (XIX) show very
rapid inversions (e.g., OH_3^+, Table 1). The greater configurational

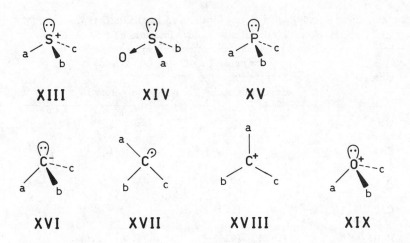

XIII   XIV   XV

XVI   XVII   XVIII   XIX

stability of second-row atoms as compared to first-row atoms is evident in Table 1.

The tricoordinate nitrogen atom has been the object of numerous studies. Fast inversion is the rule for amines (XX) (see examples in Table 1). Some substituents markedly increase the inversion barrier, which is as high as 138 kJ/mol (33 kcal/mol) for NCl_3 [4]. On the other hand, if two of the nitrogen substituents are part of a cyclic system, the barrier of inversion may, in some cases, be significantly increased. Thus, the enantiomers of 2-disubstituted aziridines (XXI) can be characterized at low temperature. Sections 10.4 and 10.5 deal extensively with nitrogen inversion in heterocycles.

When the nitrogen atom is at a ring junction in bridged ring systems (Section 10.7), pyramidal inversion is impossible without bond cleavage. Given the proper substituents, the tricoordinate nitrogen becomes a stable center of chirality, as is the case in Tröger's base (XXII).

It should be noted that most of the tricoordinate atoms discussed above bear an unshared electron pair which may be formally regarded as the

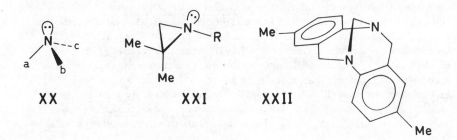

XX   XXI   XXII

fourth substituent in agreement with the pyramidal symmetry of these systems. It follows that identical stereochemical descriptors can be used for tri- and tetracoordinate chiral centers. They are presented in the following sections.

5.3 The D and L Nomenclature [5,6]

The need for consistency in stereochemical designation prompted Emil Fisher to use C-5 of the dextrorotatory enantiomer of glucose as a starting point. This molecule was drawn in the Fischer projection (see Chapter 2) after the following conventions had been fixed: (1) the longest carbon chain is vertical; (2) the most highly oxidized end of the chain is at the top; and (3) at each center along the main chain the vertical bonds point backwards. (+)-Glucose was degradated by Fischer to an aldotriose, (+)-glyceraldehyde, in which the only remaining stereogenic* center originates from C-5 of the parent molecule. Arbitrarily, Fischer assigned the configuration XXIII to (+)-glyceraldehyde, which became D-(+)-glyceraldehyde, due to the right-hand side position of the substituent on the asymmetric center.

All asymmetric molecules which could be chemically related to D-(+)-glyceraldehyde were assigned the configuration D (e.g., D-(+)-glucose), while molecules related to L-(-)-glyceraldehyde (XXIV) were assigned to the L-series.

The configurational assignment based on chemically relating asymmetric molecules to either D- or L-glyceraldehyde is known as the genetic nomenclature. It has been successfully applied to carbohydrates, but conflicting results have sometimes been recorded in other classes. Thus, D-(+)-glyceraldehyde can conceivably be chemically related to both (+)- and (-)-lactic acid [6].

To partially overcome such difficulties, the projection nomenclature came into use. This nomenclature distinguishes itself from the preceding

$$
\begin{array}{ccc}
\text{CHO} & & \text{CHO} \\
| & & | \\
\text{H}-\text{C}-\text{OH} \qquad \text{XXIII} & & \text{HO}-\text{C}-\text{H} \qquad \text{XXIV} \\
| & & | \\
\text{CH}_2\text{OH} & & \text{CH}_2\text{OH}
\end{array}
$$

---

*i.e., which introduces an element of stereoisomerism or prostereo-isomerism in a molecule.

in that no reference is made to the origin of the compound examined. The latter is simply drawn in the Fischer projection and is designated D or L, depending on the right- or left-hand position of the substituent on the asymmetric carbon or on the highest-numbered asymmetric carbon. The projection nomenclature is restricted to those molecules which can be unambiguously drawn in the Fischer projection and which can simultaneously obey all the relevant rules. Further difficulties arise for distinct chemical classes such as phenylethylamines which are conventionally drawn "upside down."

Amino acids presented further difficulties which resulted in the development of the amino acid nomenclature. While L-(-)-serine (XXV; R = CH$_2$OH) was easily designated according to the carbohydrate nomenclature, the original D-threonine (XXVI) designation is inconsistent with serine. It has been altered to L$_s$-threonine (XXVI), where the subscript s indicates the serine series; the suscript g is then used for the glyceraldehyde series.

COOH
|
H$_2$N — C — H
|
R

XXV

COOH
|
H$_2$N — C — H
|
H — C — OH
|
CH$_3$

XXVI

There are several drawbacks with the D and L nomenclature which were aptly described a few years ago [6]. These nomenclatures, as originally conceived and used, indicated nothing more than relative configurations, i.e., there was no way of deciding if all stereochemical representations reflected reality or were objects belonging to a universe enantiomeric with ours. It was not until 1951 that publications based on the X-ray analysis of sodium rubidium tartrate afforded absolute configurations and showed Fischer's gamble to have hit the correct answer, fortunately. The time was then appropriate for the appearance of a stereochemical nomenclature describing absolute configurations, and which would be simultaneously unequivocal, self-consistent, and universally applicable.

5.4 The R and S Nomenclature [5, 7, 8]

The foundation of the R and S nomenclature was laid in 1951 by Cahn and Ingold [9], and the procedure has been consolidated and markedly extended in two subsequent papers by Cahn, Ingold, and Prelog [10, 11].

The essential part of the R and S nomenclature (also called the CIP nomenclature) of chiral centers is known as the <u>sequence rule.</u> This is, in fact, a set of arbitrary but consistent rules which allow a hierarchical assignment of the substituents on the chiral center. The four substituents are designated a > b > c > d, where > denotes "is preferred to."

Conventionally, the chiral center is viewed with a, b, and c pointing toward the observer, and d pointing away. When the path a to b to c to a gives a clockwise course (XXVII), the configuration is symbolized by (R) (Latin rectus, right). When the path a to b to c to a gives a counterclockwise course (XXVIII), the symbol (S) is used (Latin sinister, left). The comparison with a spinning wheel is often made to help explain the circular path a to b to c to a.

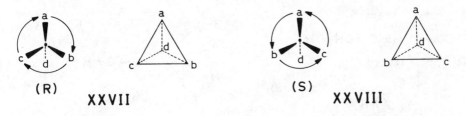

(R)   XXVII   (S)   XXVIII

The sequence rule contains five subrules which are applied, in succession, as long as necessary to reach a decision. First, the four atoms immediately adjacent to the chiral center are considered. The first subrule states that these atoms are arranged in an order of preference which decreases with decreasing atomic number. A free electron pair is considered as a phantom atom with ascribed atomic number zero. Representative atoms yield the sequence $I > Br > Cl > S > P > F > O > N > C > H$. For example, 1-bromo-1-chloroethane (XXIX) has C-1 as a chiral center; its adjacent atoms are arranged in the order $Br > Cl > C > H$ which allows clear representation of its (R)- and (S)-enantiomers.

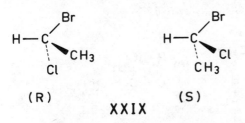

(R)   (S)

         XXIX

In the case of 2-bromo-2-chloro-3-methyl-1-butanol (XXX), two carbon atoms are adjacent to C-2, and no decision can be reached at that stage. The sequence rule states that one has to work outwards to the atoms directly attached to C-1 and C-3. One then finds C(C, C, H) and C(O, H, H), with the atoms once-removed arranged in order of preference, i.e., C(a', b', c') and C(a'', b'', c''). The atoms once-removed are compared in pairs, i.e., a' with a'', b' with b'', c' with c'', the comparison being suspended at the first difference. In our example, O > C, and the complete sequence for C-2 is Br > Cl > C(O, H, H) > C(C, C, H).

When, as in structure XXXI, the once-removed atoms C(C, C, H) and C(C, C, H) show no difference, exploration is continued further. The two sub-branches of the left-hand ligand are arranged in the order C(Cl, H, H) > C(H, H, H), and on the right-hand side we find C(O, C, H) > C(O, H, H). Comparing the senior sub-branches yields C(Cl, H, H) > C(O, C, H), and the junior sub-branches need not be compared. As a result, the left-hand ligand has preference over the right-hand ligand, and structure XXXI has (S)-configuration.

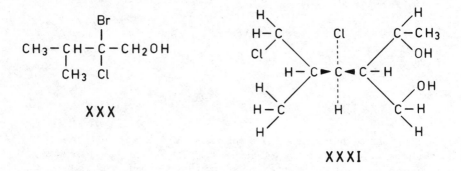

To avoid discussions on the nature of bonds, the sequence rule splits double and triple bonds into two and three single bonds, respectively. This is done by duplicating or triplicating the doubly or triply bonded atoms. The duplicated and triplicated atoms are considered as though carrying phantom atoms of atomic number zero and are drawn in brackets. Classical examples are shown in diagrams XXXII. For aromatic heterocycles, each duplicate is given an atomic number that is the mean of the

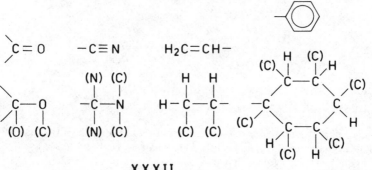

XXXII

numbers it would have if the double bonds were located at each possible position. A useful list of 76 groupings arranged according to the sequence rule can be found in the IUPAC Recommendations for Fundamental Stereochemistry [8].

The second subrule states that isotopic substituents are classified in decreasing order of mass number (e.g., T > D > H). Other subrules classify stereoisomeric ligands. The outline presented above refers to the most common cases; the original references [especially 7, 11] should be studied before considering more complex examples.

An important warning regarding the sequence rule is that chemical or biogenetic families are not necessarily correlated [7]. For example, the natural α-amino acids can be represented as structure XXV in Fischer projection, i.e., the L-configuration. Most amino acids, and among them (-)-serine (XXV; R = CH_2OH), the standard in the series, have the (S)-configuration. Cysteine (XXV; R = CH_2SH), however, is assigned the (R)-configuration due to perturbation of the sequence by the sulfur atom. Another example is seen with the (R)-alcohol (XXXIII) which yields (S)-esters. Failure of the sequence rule to consistently reveal stereochemical correlations is the reason for the survival and utility of "local" systems such as the D and L nomenclature for carbohydrates and amino acids, and the α, β nomenclature for steroids (Section 10.6).

$$CH_3-O-CH_2$$
$$\text{|}$$
$$CH_3-C-CH_2OH$$
$$\text{|}$$
$$H$$

XXXIII

References

1.   K. Mislow, Acc. Chem. Res., 3:321 (1970).

1a. K. Mislow, Pure Appl. Chem., 25:549 (1971).

1b. J. C. Martin and T. M. Balthazor, J. Am. Chem. Soc., 99:152 (1977).

2.   R. E. McMahon, H. R. Sullivan, J. C. Craig, and W. E. Pereira, Jr., Arch. Biochem. Biophys., 132:575 (1969).

3.   J. W. Henderson, Chem. Soc. Rev., 2:397 (1973).

3a. K. Ohkubo, Y. Azuma, and M. Okada, Bull. Chem. Soc. Japan, 49:1397 (1976).

3b. J. B. Lambert, Pyramidal atomic inversion, in Topics in

Stereochemistry, vol. 6 (edited by N. L. Allinger and E. L. Eliel), Wiley, New York, 1971, pp. 19-105.

4. C. H. Bushweller, W. G. Anderson, P. E. Stevenson, and J. W. O'Neil, J. Am. Chem. Soc., 97:4338 (1975).

5. W. H. Hartung and J. Andrako, J. Pharm. Sci., 50:805 (1961).

6. D. W. Slocum, D. Sugarman, and S. P. Tucker, J. Chem. Educ., 48:597 (1971).

7. R. S. Cahn, J. Chem. Educ., 41:116 (1964).

8. IUPAC Commission on Nomenclature of Organic Chemistry, 1974 recommendations for section E, fundamental stereochemistry, Pure Appl. Chem., 45:11 (1976).

9. R. S. Cahn and C. K. Ingold, J. Chem. Soc., 612 (1951).

10. R. S. Cahn, C. K. Ingold, and V. Prelog, Experientia, 12:81 (1956).

11. R. S. Cahn, C. K. Ingold, and V. Prelog, Angew. Chem. Intern. Ed., 5:385,511 (1966).

6

Stereoisomerism Resulting from
Several Centers of Chirality

When an acyclic molecule contains n nonidentical centers of chirality, it exists as 2^n stereoisomers which are enantiomeric in pairs; such a molecule exists as $2^{(n-1)}$ diastereoisomeric pairs of enantiomers. This modern version of Pasteur's rule applies only to constitutionally unsymmetrical molecules, that is, to molecules with nonidentical centers of chirality.

Constitutionally symmetrical molecules have fewer than 2^n stereoisomers and are considered separately.

6.1 Constitutionally Unsymmetrical Molecules

The simplest case (n = 2) is well-illustrated by the ephedrines (I), whose carbons 1 and 2 are asymmetric. The (1S;2R)-(+) and (1R;2S)-(-) stereoisomers share an enantiomeric relationship because they have opposed configurations on every asymmetric center; they compose the so-called erythro series, or the proper ephedrines. Similarly, the (1R;2R)-(-)- and (1S;2S)-(+)-stereoisomers are enantiomers and compose the threo series; they are also called pseudoephedrines.

Between any erythro-isomer and any threo-isomer, the relationship is not that of enantiomerism, but rather of diastereoisomerism. Indeed, two such stereoisomers have one chiral center with opposed configurations, and one with an identical configuration; they cannot be mirror images.

If we generalize the case of the ephedrines to molecules having n centers of chirality, it becomes apparent that any stereoisomer will have one enantiomer (that stereoisomer of opposed configuration on every chiral center) and 2^n-2 diastereoisomers (which may have as little as 1 and as much as n-1 centers of opposed configurations). Those diastereoisomers which differ in the configuration of a single chiral center (i.e., which have

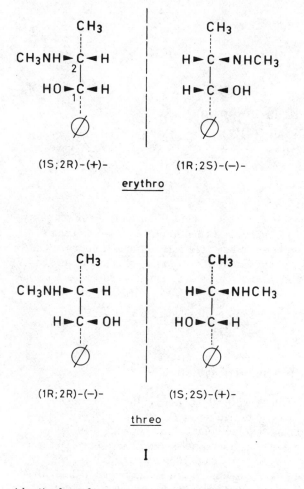

(1S;2R)-(+)-   (1R;2S)-(−)-

erythro

(1R;2R)-(−)-   (1S;2S)-(+)-

threo

I

identical configurations on n-1 centers) are called <u>epimers</u>. Any stereo-isomer will have n epimers.

Up to now, we have assumed the n asymmetric centers in a given mol-ecule to be configurationally stable. This need not always be the case, and compounds exist where one of the centers undergoes rapid inversion under normal conditions. A general case is represented by structure II, where nitrogen inversion destabilizes this chiral center and epimerizes the con-figuration of the molecule. Compound II may exist under normal conditions as two optically active forms, one a mixture of the fast interconverting epimers $R_C R_N$ and $R_C S_N$, the other of the epimers $S_C R_N$ and $S_C S_N$.

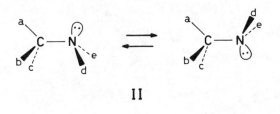

II

6.2 Constitutionally Symmetrical Molecules

Acyclic molecules having n asymmetric centers are called constitutionally symmetrical when those centers equidistant from the geometrical center of the molecule are identically substituted. Such molecules have $2^{(n-1)} + 2^{(n-2)/2}$ stereoisomers when n is even, and $2^{(n-1)}$ when n is odd.

Tartaric acid (III) is the classical example for n even. It can easily be seen that the two asymmetric carbons are identically substituted. The dextrorotatory form has the (R;R)-configuration while its levorotatory enantiomer is (S;S). The expected second pair of enantiomers (R;S) and (S;R), however, does not exist; indeed, (R;S) and (S;R) are superimposable and therefore achiral and identical. This can also be understood by considering the symmetry plane bisecting the molecule and forbidding chirality. The achiral, optically inactive stereoisomer is termed the meso-form;* it shares a diastereoisomeric relationship with the two other stereoisomers. In accordance with the above rule, tartaric acid exists as 2 + 1 stereoisomers. It is worthwhile stating explicitly that the racemic form, although physically different from the three stereoisomers (see Section 12.5), is not a stereoisomeric form.

A constitutionally symmetrical molecule having four chiral centers exists as ten stereoisomers, i.e., two meso-forms and four pairs of enantiomers.

New problems are encountered when n is odd as in trihydroxyglutaric acid (IV). The rule predicts four stereoisomers. The two stereoisomers (S;S) and (R;R) differ in the configuration of C-2 and C-4. In these two

---

*The use of "meso" in characterizing some stereoisomeric forms is not to be confused with the "meso-carbon atom." The latter, which designates carbon atoms of the type C(aabc) [1], is now superseded by the concept of the "prochiral center" to be discussed later (see Section 11.2); the central carbon of glycerol $HOCH_2$-$CH(OH)$-CH_2OH is such a prochiral center.

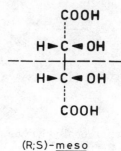

```
      COOH              COOH               COOH
                 |
  H►C◄OH      |   HO►C◄H          H►C◄OH
                 |                    ─ ─ ─ ─ │ ─ ─ ─ ─
 HO►C◄H      |    H►C◄OH          H►C◄OH
                 |
      COOH      |      COOH               COOH

  (R;R)-(+)       (S;S)-(−)          (R;S)-meso
```

III

```
      COOH       |       COOH
                 |
 HO►C◄H       |    H►C◄OH
                 |
  H►C◄OH      |   HO►C◄H
                 |
  H►C◄OH      |   HO►C◄H
                 |
      COOH      |       COOH

    (S;S)          |      (R;R)
```

IV

```
      COOH               COOH

  H►C◄OH             H►C◄OH

  H►C◄OH            HO►C◄H

  H►C◄OH             H►C◄OH

      COOH               COOH

 (R;r;S)-meso       (R;s;S)-meso
```

stereoisomers, the central C-3 carries two identical substituents which are the (S)-glycolyl moieties in the case of the (S;S)-stereoisomer, and the (R)-glycolyl moieties in the other stereoisomer. Thus, carbon-3 is achiral, and, more precisely, it is a prochiral center (Caabc, see Section 11.2)[2]. The (S;S)- and (R;R)-stereoisomers have only two centers of chirality, and since the stereoisomers have opposed configurations at both of these centers, they are enantiomers (optically active; see Section 12.3).

When in trihydroxyglutaric acid the carbon-2 and -4 have opposed configurations, the molecule has a plane of symmetry perpendicular to its main axis and cutting C-3; the molecule is achiral. In this case, however, C-3 has four different substituents, namely -H, -OH, an (R)-glycolyl moiety and an (S)-glycolyl moiety (the latter two substituents are enantiomorphic; see Section 4.2). Because of this situation, C-3 may have two opposed configurations, allowing the achiral molecule to exist in two distinct, optically inactive stereoisomeric forms, both called <u>meso</u>-forms. We now face the particular situation of a carbon having four different substituents but lying in a plane of symmetry of the molecule. Such a carbon is referred to as a <u>pseudoasymmetric atom</u>. Its general expression is Ca^+a^-bc [2,3], with a^+ and a^- symbolizing the two enantiomorphic ligands.

A subrule of the sequence rule states that R > S and allows pseudoasymmetric centers to be treated in the same way as chiral centers. However, since the molecule is achiral, these centers are given the lowercase symbol r or s.

In compounds such as IV where n is odd, there are always (n-1) asymmetric centers and one pseudoasymmetric atom. Among the $2^{(n-1)}$ stereoisomers, every <u>meso</u>-form is diastereoisomeric with all other $2^{(n-1)}-1$ isomers, while the chiral stereoisomers are enantiomers with one isomer and diastereoisomeric with the $2^{(n-1)}-2$ others.

Another representation of a pseudoasymmetric center is as in structure V; Va and Vb built a pseudoasymmetric pair (e.g., the two <u>meso</u>-forms of structure III) [4]. Reflection of a <u>meso</u>-form inverts the enantiomorphic groups F, but leaves the pseudoasymmetric center and the molecule itself with unaltered absolute configuration.

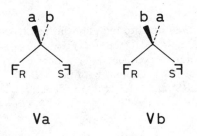

Va   Vb

6.3 Extension of the R and S Nomenclature to Designate Relative Configurations

The application of the R and S nomenclature to designate the absolute configuration of molecules with several centers of chirality is straightforward, as illustrated without comments at the beginning of the chapter. When studying molecules with several centers of chirality, the chemist not infrequently faces the problem of designating one pure enantiomer of known relative, but unknown absolute configuration. For such cases, the R and S nomenclature has been expanded, the prefixes R* and S* (R-star and S-star) being adopted to describe relative configuration.

The assignment of the R* and S* prefixes by the sequence rule is based on the arbitrary assumption that the center of chirality with the lowest locant has the (R)-chirality [5]. Take for example the bromo-chloro-nitrocyclohexane derivative (VI); the senior substituent is the bromo group, and the second senior substituent is the chlorine atom. The lowest locant is attributed to the carbon atom carrying the senior substituent, and structure VI is, therefore, 1-bromo-3-chloro-5-nitrocyclohexane. Let us assume structure VI to be one pure enantiomer of unknown absolute, but known relative configuration as shown (R,S,R or S,R,S). As indicated above, C-1 is arbitrarily attributed the (R)-configuration, from which it follows that the relative configuration of structure VI is (1R*;3S*;5R*). Alternatively, the prefix rel (for relative) may be used, i.e., rel-(1R;3S;5R)-1-bromo-3-chloro-5-nitrocyclohexane [5].

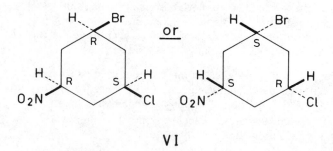

VI

The R* and S* nomenclature is sometimes used to indicate the relative configuration of racemates [see the two examples (1R*;2R*)-(±) and (1R*;2S*)-(±)] rather than that of optically active compounds of unknown absolute configuration. This practice is not in agreement with the IUPAC rules [5], which recommend writing (1RS;2RS) and (1RS;2SR), respectively.

References

1. P. Schwartz and H. E. Carter, Proc. Natl. Acad. Sci. , USA, 40:499, (1954).

2. H. Hirschmann and K. R. Hanson, Tetrahedron, 30:3649 (1974).

3. V. Prelong and G. Helmchen, Helv. Chim, Acta, 55:2581 (1972).

4. J. G. Nourse, J. Am. Chem. Soc. , 97:4594 (1975).

5. IUPAC Commission on Nomenclature of Organic Chemistry, 1974 recommendations for section E, fundamental stereochemistry, Pure Appl. Chem. , 45:11 (1976).

7

Stereoisomerism Resulting from Axes and Planes of Chirality, and from Helicity

The presence of a center of chirality is not a necessary condition for molecular dissymmetry. According to the factorization rule [1], the over-all chirality can be factorized into three elements, chiral centers, chiral axes, and chiral planes; whenever necessary, they are treated in this order. Another element of chirality can be gainfully considered, namely, helicity. This chapter is concerned with the nomenclature and configurational aspects [2-4] of molecules displaying axial, planar, and helical dissymmetry. An extensive review of the topic, published a few years ago, may also be consulted [5].

7.1 The Chiral Axis

When the four substituents a, b, c, d, of a central atom X occupy the vertices of a tetrahedron (I), an asymmetric center exists in the molecule. If the tetrahedron is elongated as in structure II, the chiral center X is extended to a chiral axis XY. In such an elongated tetrahedron, the conditions required for chirality are less stringent than in a regular tetrahedron. Indeed, structure III shows that the minimal condition for chirality is that a be different from b. In other words, an elongated tetrahedron will be chiral if the pair of substituents at the X end of the axis and the pair at the Y end are each formed by two different substituents.

When the sequence rule is to be applied to axial chirality, an additional rule is needed which states that the two near groups precede the two far groups. By so doing, the (R) or (S) designation of a chiral axis becomes independent from the way the axis is viewed. Thus, viewing structure II from the X end, and assuming a>b and c>d, yields the sequence shown in model IVa, whereas viewing structure II from the Y end yields model IVb. As in the usual procedure, viewing the two models IVa and IVb from the side remote from 4 results in a clockwise path 1-2-3. Models IVa and IVb

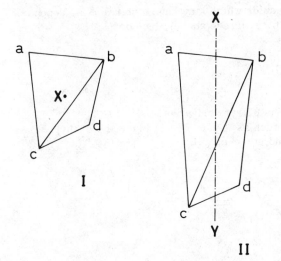

I

II

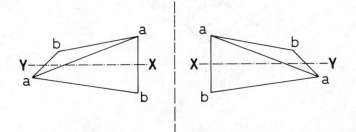

III

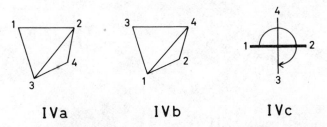

IVa   IVb   IVc

are identical, and structure II is assigned the (R)-configuration. Another way to present structure II is to dispose horizontally the two near groups, and vertically the two far groups: A horizontal and a vertical line join each pair; they are the two relevant edges of the tetrahedron. Diagram IVc is obtained, in which the path from 1 to 2 to 3 is clockwise, i.e., (R).

Selected examples of molecules which may have a chiral axis, depend-ing on the nature of groups a, b, c, d (see above), include

Allenes (V)
Alkylidenecycloalkanes (VI)
Spiranes (VII)
Adamantanes (VIII)
Biphenyls (IX) and analogous biaryl derivatives
Singly bridged biphenyls, such as X, and doubly bridged biphenyls
Substituted anilines (XI) and styrenes (XII)

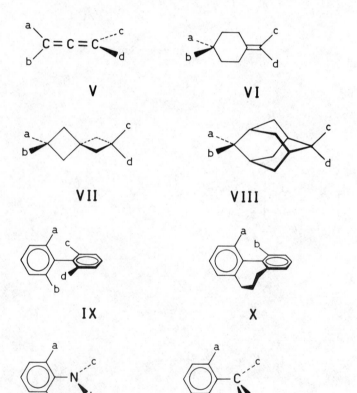

Many of the above axially chiral molecules can have a C_2 axis as the sole element of symmetry and, therefore, belong to point group C_2. This is the case when a = c and b = d for allenes (V) (see Chapter 1, dia-gram IX), alkylidenecycloalkanes (VI), spiranes (VII), adamantanes (VIII), and biphenyls (IX). This is easily tested by drawing the molecules in diagrams similar to structure IXb in Chapter 1.

Let us now discuss the above groups of axially chiral molecules and give some examples. Structures V and VI resort to torsional isomerism about double bonds and will be discussed in this context in Section 8.1. An example of a chiral allene is (R)-(-)-glutinic acid (XIII) [5].

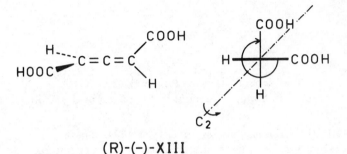

(R)-(-)-XIII

Spiranes such as structure VII and adamantanes (VIII) owe their possible axial chirality to their particular cyclic skeleton. Recently, (+)-spiro [3.3] heptane-2,6-dicarboxylic acid (Fecht acid) has been proven to have the absolute configuration shown in diagram XIV, i.e., the (S)-configuration [6].

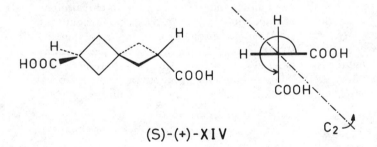

(S)-(+)-XIV

The enantiomers of spiro [4.4] nonane-1,6-dione have been synthesized and their absolute configurations determined (see Ref. 5); the levorotatory enantiomer is shown in diagram XV. If this molecule were considered to be axially chiral, then its configuration would be (R). However, spiranes such as in structures XV and XVI (but not XIV) are now recognized to be centrally chiral. Indeed, besides the centrally chiral molecules lacking any element of symmetry (group C_1; see Section 5.1), molecules having one twofold axis of symmetry (C_2) (see XV and above) can contain chiral centers of stereoisomerism [4,5] whose configuration can be clearly defined. In the case of structure XV, the rule states that the sequence

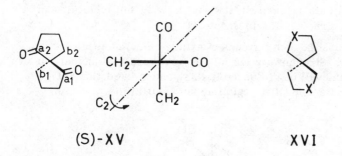

(S)-XV   XVI

starts with either carbonyl, and proceeds in the order a_1-a_2-b_1-b_2 (alternatively a_2-a_1-b_2-b_1); both sequences yield the (S)-configuration. The
interested reader is referred to a major publication [4] for an in-depth discussion of C_2 chiral centers.

Compounds IX–XII are torsional isomers about single bonds; their
barrier of torsion varies widely with the substituents. Some derivatives
are stable enough to render resolution feasible, while others are practically
unresolvable. Isomers resulting from restricted rotation about single
bonds are called atropisomers. Their conformational aspects and energies
of isomerization will be discussed in the general context of torsional isomerism about single bonds (Chapter 9).

In the present chapter, the discussion of atropisomers is restricted to
nomenclature problems, in particular, those encountered in chiral biphenyls.
In the case of 2,6,2',6'-substituted biphenyls (e.g., IX), no difficulties are
encountered in applying the usual rules. In the case of more complex substitution patterns (e.g., XVII), the ring carbons 2, 6, 2', 6', are again
considered; structure XVII has the (R)-configuration [1]. Note also that
structure XVII has no C_2 and is asymmetric. Biaryls in general, as well
as 2,2'-bridged biphenyls and biaryls, are treated in a comparable manner [5,7].

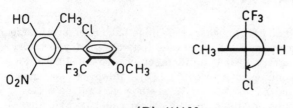

(R)-XVII

In the case of molecules possessing several elements of chirality, the
designation of the absolute configuration of the chiral axis as (R) and (S)
may lead to insufficient clarity. For such molecules the prefixes (aR) and

(aS) are used, indicating that the stereochemical descriptor refers to a chiral axis.

7.2 The Chiral Plane

In a compound such as XVIII, the plane containing the benzene ring, the two oxygen atoms, and the bromine atom is called a plane of chirality. The molecule is chiral without displaying any center or axis of chirality. Structure XVIII represents one enantiomer, the other enantiomeric form having the bridge on the opposite side of the chiral plane. The plane containing the same atoms in the analogous molecule lacking the polymethylene bridge (2-bromohydroquinone) is a plane of symmetry. Incorporation of the bridge into 2-bromohydroquinone "desymmetrizes" the plane and transforms it into a chiral plane.

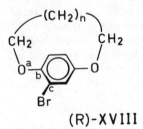

(R)-XVIII

The sequence rule has been extended to include cases of stereoisomerism involving a chiral plane. The sequence rule-preferred atom directly attached to the plane is chosen as the pilot atom. In structure XVIII, this is the carbon atom of the left-hand methylene group. Starting from the pilot atom, one classifies the atoms of the plane in their order of encounter along the bonds. When an ambiguity arises, the sequence rule is applied. Thus, the path explored from the pilot atom is the sequence rule-preferred path, and this is done until a clockwise (R) or counterclockwise (S) rotation is traced when viewing the molecule from the pilot atom. In structure XVIII, this path is O-C-C(Br), and it is clockwise (R). For the sake of clarity, the prefixes (pR) and (pS) may be used to indicate that the configuration refers to planar chirality.

A more complex case of planar chirality is represented by paracyclophanes such as the carboxylic acid derivatives of structure XIX. [2.2]Paracyclophanecarboxylic acid (m = n = 2) has been resolved; it is optically stable, and the dextrorotatory enantiomer has been assigned the (S)-configuration (XX) [5]. In this molecule, the plane of chirality must be defined; it is the plane of the carboxylic-substituted ring [1], and the pilot atom is underlined. Among the higher homologs, (m = 3, n = 4) is still resolvable, but racemizes at high temperature, while (m = n = 4) cannot be resolved.

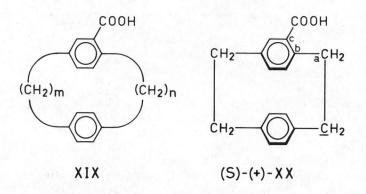

XIX   (S)-(+)-XX

Another case of planar chirality is provided by trans-cycloalkenes as exemplified by trans-cyclooctene (XXI). In this molecule, the chiral plane contains the two double-bonded carbon atoms, as well as the two hydrogens and the two carbons directly adjacent. Two equivalent pilot atoms exist (underlined in XXI); in the absolute configuration drawn in structure XXI, the two pilot atoms each yield the (S)-configuration because the two paths are C-C=C. The enantiomers of trans-cyclooctene have been resolved and their absolute configuration determined as (R)-(-) and (S)-(+); they are represented in diagram XXII [5].

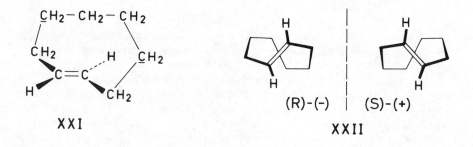

XXI   (R)-(-) | (S)-(+)
                                      XXII

7.3 Helicity

Helices (cylindrical, conical) are chiral objects (XXIII). A right-handed helix (which shows a clockwise rotation when viewing it along the axis and moving from the front to the rear) is designated P (plus) while a left-handed helix is M (minus).

Helical chirality is often encountered in nature, for example, in helical shells and in whirlpools. The corkscrew is a famous example of a manufactured object displaying helicity. Chemistry also knows helical chirality, which is encountered in rigid molecules (configurational helices) and in flexible molecules (conformational helices). Helical molecules have axial dissymmetry, being a particular category of structures with a chiral axis. However, the concept of helicity affords a straightforward and very convenient tool for comprehending and designating their chirality.

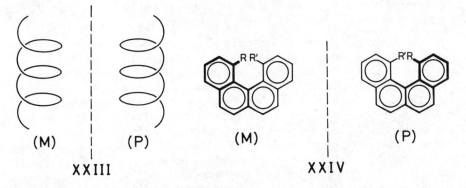

(M)   (P)   (M)   (P)

XXIII   **XXIV**

Steric crowding in benzophenanthrenes such as compounds XXIV prevents the two substituents from being in the same plane and twists the whole molecule out of planarity. Of great interest to chemists is the class of molecules known as helicenes, as exemplified by hexahelicene (XXV). This molecule has been resolved[8] and its absolute configuration determined (see Ref. 5 for review); the dextrorotatory enantiomer has (P)-helicity (XXV). Hexahelicene has an inherently dissymmetric chromophore (see Sections 12.3 and 12.6) and exhibits spectacularly high optical activity. These properties have been a stimulus to research, and higher homologs such as octahelicene and nonahelicene have been synthesized and studied (e.g., Ref. 9). Other interesting models are the heterohelicenes (e.g., Ref. 10), helicenes in which one or several benzo rings are replaced by heterocycles.

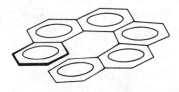

(P)-(+)-**XXV**

Important and famous examples of conformational helicity are provided by some secondary structures of polypeptides. Many proteins have significant portions of their chains stabilized in the α-helix, which can be either right- or left-handed. This helix results from intrachain hydrogen bonds between N-H and C=O groups four residues apart; there are 3.7 amino acid residues per turn, and the pitch is 5.4 Å. A less frequent form has been characterized, namely the ω-helix, a distorted form of the α-helix. There are four residues per turn, and the pitch is 5.3 Å. The interested reader may consult standard textbooks in biochemistry, and a thorough review on polypeptide stereochemistry for more information [11].

As mentioned above, helical molecules are axially chiral, but their configuration is not conveniently described in terms of the (aR) and (aS) nomenclature. On the other hand, it may be useful in some cases to describe in terms of the helical nomenclature the configuration or confor- mation of axially chiral molecules whose helicity is not superficially obvious. For example, chiral conformations of ethane derivatives can be designated by the helicity rule. The sequence rule-preferred ligands are considered as shown in the general examples XXVI. Similarly, the stereo- chemistry of biphenyls can be described in terms of helicity. This applica- tion is particularly useful in the case of unresolvable biphenyls such as those in diagram XXVII, in which a configurationally stable chiral axis

(M)

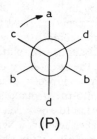

(P)

XXVI

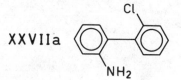

XXVIIa

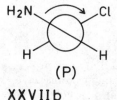

(P)

XXVIIb

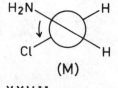

(M)

XXVIIc

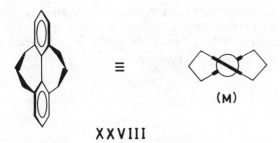

XXVIII

does not exist, and for which the use of the essentially configurational descriptors R and S would be inadequate. On the other hand, configurationally stable singly and doubly bridged biphenyls (e.g., XXVIII) are sometimes usefully characterized by their helicity. Interesting applications to aporphine alkaloids have been reported [12].

References

1. R. S. Cahn, C. Ingold, and V. Prelog, Angew. Chem. Intern. Ed., 5:385,511 (1966).

2. R. S. Cahn, J. Chem. Educ., 41:116 (1964).

3. IUPAC Commission on Nomenclature of Organic Chemistry, 1974 recommendations for section E, fundamental stereochemistry, Pure Appl. Chem., 45:11 (1976).

4. H. Hirschmann and K. R. Hanson, J. Org. Chem., 36:3293 (1971).

5. G. Krow, The determination of absolute configuration of planar and axially dissymmetric molecules, in Topics in Stereochemistry, vol. 5 (edited by E. L. Eliel and N. L. Allinger), Wiley, New York, 1970, pp. 31-68.

6. L. A. Hulshoff, H. Wynberg, B. van Dijk, and J. L. de Boer, J. Am. Chem. Soc., 98:2733 (1976).

7. D. M. Hall, The stereochemistry of 2,2'-bridged biphenyls, in Progress in Stereochemistry, vol. 4 (edited by B. J. Aylett and M. M. Harris), Butterworths, London, 1969, pp. 1-42.

8. M. S. Newman and D. Lednicer, J. Am. Chem. Soc., 78:4765 (1956).

9. R. H. Martin, M. Flammang-Barbieux, J. P. Cosyn, and M. Gelbcke, Tetrahedron Lett., 3507 (1968).

10. H. Wynberg and M. B. Groen, J. Am. Chem. Soc., 90:5339 (1968).

11. M. Goodman, A. S. Verdini, N. S. Choi, and Y. Masuda, Peptide stereochemistry, in Topics in Stereochemistry, vol. 5 (edited by E. L. Eliel and N. L. Allinger), Wiley, New York, 1970, pp. 69-166.

12. G. G. Lyle, J. Org. Chem., 41:850 (1976).

8

Torsional Isomerism About Double Bonds

Torsional isomerism about double bonds will be discussed in this chapter, while single bonds constitute the subject matter of Chapter 9. No isomerism can result from triply bonded tetravalent or trivalent atoms, since such systems display cylindrical symmetry (Chapter 1). Torsional isomerism about triply bonded pentavalent and hexavalent centers is for future generations of chemists to consider.

8.1 π Diastereoisomerism About Carbon-Carbon Double Bonds

The coplanarity of the six atoms in a structure such as I is a well-known fact, and arises from factors presented in Section 3.3. The molecule I is achiral, but it has a stereoisomeric form (II) with which it shares a diastereoisomeric relationship. Because compounds I and II have identical substituents on the same and on opposite sides of the double bond, respectively, they are termed <u>cis</u> and <u>trans.</u> Their stereoisomeric relationship is often called <u>cis-trans</u>-isomerism. The designation π diastereoisomerism [1] appears more appropriate since it conveys the chemical origin and the correct designation of the stereoisomeric relationship; this designation is further useful in that it avoids any confusion with <u>cis-trans</u>-isomerism in a cyclic system where no double bond is involved (σ diastereoisomerism; see Section 10.3).

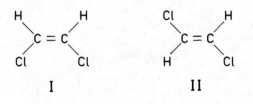

I   II

The designation of stereoisomers I and II as cis and trans, respectively, presents no difficulty. In the general case of structure III, the minimum condition for π diastereoisomerism requires a ≠ b and c ≠ d. Until recently, the designation of structure III as cis or trans was made with respect to the largest or more remarkable groups, or with respect to the chain continuing on both sides of the double bond. For example, angelic acid (IV) and tiglic acid (V) have long been considered as the trans- and cis-forms of 2-methyl-2-butenoic acid, respectively, based on the relative positions of the two methyl groups.

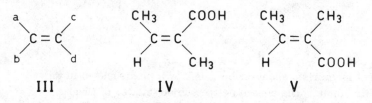

III   IV   V

It is obvious that such a nomenclature is ambiguous and relies on inconsistent rules. To overcome this difficulty, it was proposed to separately classify the pairs (a, b) and (c, d) according to the sequence rule, and to designate seqcis and seqtrans the configurations having the two sequence rule-preferred substituents in a cis- and trans-position, respectively. The prefixes seqcis and seqtrans, however, were rapidly superseded by the prefixes Z (German zusammen = together) and E (entgegen = opposite), respectively [2]. Therefore, and according to IUPAC rules [3], the prefixes Z and E mean that the sequence rule-preferred substituents in each pair are on the same or opposite sides of the C=C axis, respectively. While in most cases Z is equivalent to the conventional cis, and E to trans, this may not always be the case. For example, compounds IV and V are now called (Z)- and (E)-2-methyl-2-butenoic acid, respectively [3].

Diastereoisomerism about carbon-carbon double bonds resorts to configurational isomerism because the isomers are separated by a high-energy barrier. On the other hand, the cis- and trans-isomers do not have the same stability since their energy contents are different. We shall consider first the isomerization process, then briefly discuss the relative energies of the two isomers.

Figure 1 represents the general curve of torsional strain variation in olefins with varying torsion angle [4,5]. The two low-energy planar states are (E) and (Z), while the two transition states have (M)- and (P)-helicity, respectively.

The conversion of one isomer into another, as shown in Figure 1, occurs by rotation about the double bond, whereby the π component of the double bond is formally cleaved. Theoretical calculations of the ethylene

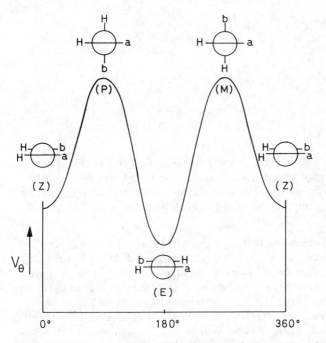

FIGURE 1 General form of the torsional strain V_θ about double bonds. From Mislow [4] and Kalinowski and Kessler [5].

rotation barrier yielded results between 226 and 428 kJ/mol (54 and 102 kcal/mol), with 344 kJ/mol (82 kcal/mol) as the best value. Experimentally, the energy barrier of (E)-1,2-dideuterioethylene was found to be 272 kJ/mol (65 kcal/mol) [5]. Steric, and mainly electronic, contributions of substituents greatly influence the barrier height. As a general trend, a lowering in bond energy results in a lower barrier [5]. For example, the barrier for (Z)-1,2-diphenylethylene (III; a = c = phenyl; b = d = H) is about 180 kJ/mol (43 kcal/mol) [5]. Derivatives with an electron-withdrawing substituent on one carbon and an electron-donating substituent on the other show diminished bond orders by charge separation consequently, they experience greatly facilitated rotation [5]. Rotation values down to 40 kJ/mol (about 10 kcal/mol) are reported; these imply conformational isomerism.

Rotation, as discussed up to now, means motion of substituents in a plane perpendicular to the C=C axis (VI). Alternative pathways for isomerization exist and the most significant is inversion [5,6]. In the latter case, isomerization occurs with motion of a substituent within the molecular plane (e.g., VII). In the case of carbon-carbon double bonds,

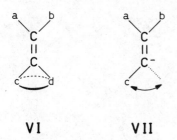

VI   **VII**

inversion must be preceded by the homolytic or heterolytic removal of a substituent; this cleavage of a σ bond is obviously a high-energy process, while the inversion process itself requires a few kcal/mol for a carbon radical, and approximately 125 kJ/mol (30 kcal/mol) for a carbanion (VII, a = b = c = H) [5,6].

   Isomerization experiments lead to an equilibrium ratio of the cis- and trans-isomers, which is a function of their relative stability. Several through-space and through-bond interactions discussed in Section 3.3 contribute to the relative stability of the two diastereoisomers. It is generally found that steric effects favor the more extended (i.e., trans) form by 4 to 40 kJ/mol. For example, trans-2-butene (VIII) is more stable than the cis-isomer by 4 kJ/mol [4,7]. Through-bond and through-space attractive orbital interactions have, however, been calculated in many cases to stabilize the cis-isomer. Thus, cis-1,2-dichloroethene (I) and cis-1-methoxypropene (IX) are more stable than their trans-isomers by about 2 kJ/mol [4,7]. In compound IX, interactions between orbitals of the methyl group and of the heteroatom have been shown to be operative [7].

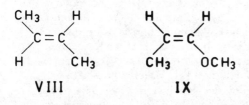

VIII   **IX**

8.2 Torsional Chirality in Carbon-Carbon Double Bonds

   The previous section discusses systems containing one carbon-carbon double bond; the four substituents lie in the same plane, and the C=C axis is an axis of diastereoisomerism. On the other hand, the allenes (X) presented in Section 7.1 are C-C double bond systems in which the four substituents lie in two perpendicular planes; an axis of chirality may exist in allenes depending on the nature of four substituents a, b, c, d (X).

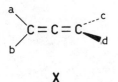

X

The stereochemical aspects of olefins and allenes can be discussed and generalized in the context of cumulenes. These molecules are made up of a chain of two or more cumulated carbon-carbon double bonds, and carry four substituents a, b, c, d. The condition for stereoisomerism is $a \neq b$ and $c \neq d$. Further, if the four substituents lie in the same plane, the molecule possesses an axis of diastereoisomerism; this is encountered in cumulenes with an odd number of double bonds (e.g., XI). But when the number n of double bonds is even, the four substituents lie in two different planes and a chiral axis exists.

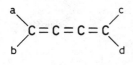

XI

Thus, it appears that once the condition $a \neq b$ and $c \neq d$ is fulfilled, the axis of stereoisomerism in cumulenes will be an axis of diastereoisomerism or of enantiomerism, depending on the coplanarity or noncoplanarity of a and b versus c and d.

As a digression, let us state that ethylene can be viewed as a "cumulene" with n = 1; π diastereoisomerism requires 1, 3, 5 ... "cumulated" double bonds. As pointed out by one of the author's students, the case n = 0 is, in fact, the asymmetric carbon itself, assuming $a \neq b \neq c \neq d$ and not simply $a \neq b$ and $c \neq d$. It follows that enantiomerism may occur for "cumulenes" with n = 0, 2, 4 ..., and that chemistry suggests zero to be an even number!

The cumulenes interest chemists not only in the static configurational aspects just discussed, but also in the dynamic and mechanistic aspects of isomerization (e.g., Ref. 8). A major merit of these molecules is to provide a system in which diastereoisomerism and enantiomerism show unusually close analogies.

Noncoplanarity of the four substituents a, b, c, d, discussed above, is also found in the class of molecules known as the twisted olefins. Consider, for example, the recently reported stereoisomers of 4,4'-bi-1,1', 2,2',3,3'-hexahydrophenanthrylidene (XII) [9]. Two diastereoisomers have been synthesized, namely, the cis- and the trans-isomer (XIIa and

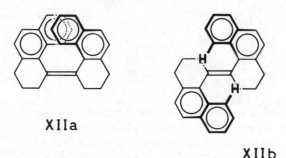

XIIa

XIIb

XIIb, respectively). Both diastereoisomers are sterically overcrowded and, as a consequence of the distorsion and nonplanarity about the olefinic bond, the axis along this bond becomes an axis of chirality. This is proven by the resolution of the two diastereoisomers into their enantiomers [9]. The configuration of the enantiomers is best specified in terms of helicity, since structure XIIa is a simple helix, and structure XIIb a double helix. However, the absolute configuration of the resolved enantiomers is not known at present.

Several twisted olefins exist which would remain chiral even if the distorted π bond could be forced into planarity. This means that chirality resulting from torsion is not the only element of chirality in the molecule. Consider trans-cyclooctene (Section 7.2, structure XXII); we have discussed this molecule in terms of its plane of chirality. It is known, however, that a substantial π bond torsion exists in this molecule [10]. This additional element of chirality cannot vary independently from the configuration about the plane of chirality; (R)-(-)-trans-cyclooctene has (P)-helicity along the olefinic bond, but cannot have (M)-helicity. Therefore, and despite the presence of two elements of chirality, trans-cyclooctene can exist only as two enantiomers, and not as four stereoisomers.

Another interesting twisted olefin is twistene (XIII), whose enantiomers have been synthesized and their absolute configuration determined as shown [11,12]. The (+)-enantiomer has (R)-configuration on the four asymmetric carbons and (P)-helicity when viewing the molecule along the C_2 axis, and also along the olefinic and CH_2-CH_2 bonds.

8.3 π Diastereoisomerism About C-N and N-N Double Bonds

Imino derivatives (XIV) differ from ethylene derivatives (VI) in that the inversion process does not require bond cleavage. Quantum calculations predict for methyleneimine (XIV; a = b = c = H) a barrier of rotation of about 250 kJ/mol (60 kcal/mol), and a barrier of inversion of about 125 kJ/mol (about 30 kcal/mol) [5]. It is clear in this simple case that

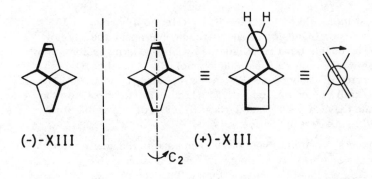

(-)-XIII   (+)-XIII

C_2

XIV

inversion is strongly favored over rotation, with a lowered barrier of overall isomerization as compared to ethylenes.

However, the effects of substituents on both processes may be different, and any of the two mechanisms could conceivably prevail in selected cases. An intermediate mechanism has even been proposed for some imines [5].

The isomerization barrier is quite sensitive to steric, electronic, and solvent effects; for many imines, it lies in the range of 80 to 130 kJ/mol (about 20 to 30 kcal/mol). Electronegative substituents on the nitrogen atom increase stability toward inversion, and this effect is seen in the relatively good stability of oximes (XIV; c = OH) and hydrazones (XIV; c = NRR').

Isomerization at a nitrogen-nitrogen double bond (azo derivatives; XV) occurs by inversion at one of the nitrogen atoms. Quantum calculations

XV

for the simplest molecule (XV; a = b = H) have led to a barrier of 193 kJ/ mol (46 kcal/mol) for inversion at one nitrogen, but about 300 kJ/mol (70 kcal/mol) for double bond rotation; the synchronous inversion of both nitrogen atoms would require about 400 kJ/mol [5]. The observed iso- merization barrier of azobenzene (XV; a = b = phenyl) is about 96 kJ/mol (about 23 kcal/mol), as compared with 75 kJ/mol (18 kcal/mol) for the analogous imine (XIV; a = c = phenyl; b = H) and with the value of the analogous olefin reported in Section 8.1.

The prefixes <u>syn</u> and <u>anti</u> have often been used to designate the π diastereoisomers resulting from the presence of a stereogenic C-N or N-N double bond. As previously discussed for olefins, the prefixes Z and E are now in general use and eliminate ambiguities formerly encountered. For example, compound XVI is called (Z, E)-(benzil dioxime) [3].

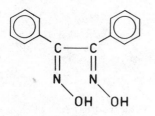

XVI

Normally, the (E)-isomers of imino and azo derivatives are thermo- dynamically preferred over the (Z)-isomers. Many factors such as resonance stabilization, steric repulsions, and nonbonded attractive or repulsive interactions influence the E:Z ratio at equilibrium [5,13,14].

References

1. J. L. Pierre, *Principes de Stéréochimie Organique Statique*, Armand Colin, Paris, 1971, pp. 103-108.

2. J. E. Blackwood, C. L. Gladys, K. L. Loening, A. E. Petrarca, and J. E. Rush, *J. Am. Chem. Soc.*, 90:509 (1968).

3. IUPAC Commission on Nomenclature of Organic Chemistry, 1974 recommendations for section E, fundamental stereochemistry, *Pure Appl. Chem.*, 45:11 (1976).

4. K. Mislow, *Introduction to Stereochemistry*, Benjamin, New York, 1966, pp. 38-39.

5. H. O. Kalinowski and H. Kessler, Fast isomerizations about double bonds, in Topics in Stereochemistry, vol. 7 (edited by N. L. Allinger and E. L. Eliel), Wiley, New York, 1973, pp. 295-383.

6. H. Kessler, Tetrahedron, 30:1861 (1974).

7. N. D. Epiotis, D. Bjorkquist, L. Bjorkquist, and S. Sarkanen, J. Am. Chem. Soc., 95:7558 (1973).

8. K. Bertsch, G. Karich, and J. C. Jochims, Chem. Ber., 110:3304 (1977).

9. B. Feringa and H. Wynberg, J. Am. Chem. Soc., 99:602 (1977).

10. A. F. Drake and S. F. Mason, Tetrahedron, 33:937 (1977).

11. M. Tichý and J. Sicher, Tetrahedron Lett., 4609 (1969).

12. M. Tichý, Tetrahedron Lett., 2001 (1972).

13. J. Bjørgo, D. R. Boyd, C. G. Watson, and W. B. Jennings, J. Chem. Soc. P.T. II, 757 (1974).

14. J. Bjørgo, D. R. Boyd, D. C. Neill, and W. B. Jennings, J. Chem. Soc. P.T. I, 254 (1977).

9

Torsional Isomerism About Single Bonds

9.1 Thermodynamic Aspects of Conformation

This chapter is devoted mainly to conformational isomerism. Indeed, isomers generated by rotation about a single bond are, in most cases, separated by an energy barrier of several kilojoules per mole, which corresponds to extremely rapid rates of interconversion.*

The energy differences between rotational isomers (rotamers) are relatively small, on the order of several kilojoules per mole; it is this energy difference which is the key factor in determining the isomeric composition of a flexible compound. In turn, this conformational composition may be a major factor in influencing the reactivity, especially the biological reactivity, of a given compound. The conformational free energy difference $\Delta G°$ for the case A $\rightleftharpoons$ B is related to the conformational equilibrium constant K (or conformational ratio) by the equation:

$$-\Delta G° = RT \cdot \ln K \tag{1}$$

Note that this equation applies to all dynamic equilibria, not only conformational ones. If K or $\Delta G°$ is known, the percentage of the more stable conformer at the given temperature can be calculated. Figure 1 plots percent of more stable isomer versus free energy difference of isomers at two temperatures. It can be seen that energy differences of 4.2 and 12.6 kJ/mol (1 and 3 kcal/mol) correspond to isomeric compositions of 85:15 and 99.5:0.5, respectively. Comprehensive tabulations covering a wide range of temperatures have been published [1].

---

*From Eyring equation (Chapter 4, Eq. 1), it can be calculated that barriers of 20, 40, and 60 kJ/mol (5, 10, and 15 kcal/mol) correspond approximately to rate constants of interconversion of 10^9, 10^5, and 10^2 sec^{-1}, respectively (see also footnote p. 37).

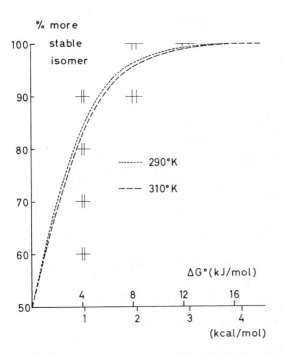

FIGURE 1 Isomeric composition (case A ⇋ B) as a function of the energy difference between isomers A and B.

The concept of conformation germinated in the 1940s owing to the creative contributions of Hassel [2,4] and Barton [3,4]. Despite this relatively recent appearance, spectacular progress has been recorded and continues to be reported at an exponential rate. Two classical books which are important milestones in the development and evolution of the conformation concept can be found in Refs. 5 and 6.

9.2 Unconjugated Systems

In this section, we shall consider rotation about single bonds linking unconjugated centers, or centers which can be regarded as unconjugated (sp^3- sp^3 carbon-carbon bonds, sp^3 carbon-heteroatom bonds, and heteroatom-heteroatom bonds).

The classical way to start a discussion on conformation is with ethane, a simple, but rich enough model. Internal rotation of this simple rotor is best seen using Newman projection (I). Assuming the front carbon atom and its three hydrogens to be stationary, the rear atoms are left free to

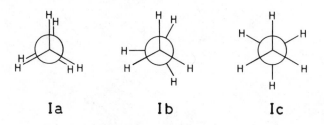

| Ia | Ib | Ic |

rotate. Since the torsion angle can assume any value, an infinite number of rotational isomers of ethane are conceivable. Two of them are remarkable, structure Ia, which displays an eclipsed conformation, and the staggered conformer in structure Ic. In structure Ia, the torsion angle (better designated dihedral angle) is 0°, while in structure Ic it is 60°. All intermediate conformations are called skewed (e.g., Ib).

The staggered conformations are the low-energy forms, while the eclipsed conformations represent transition states. The energy difference between the two conformers, which, in fact, represents the barrier of interconversion between two staggered conformers, is very close to 12 kJ/mol (about 3 kcal/mol). A complete rotation of 360° generates three identical staggered and three eclipsed forms (Figure 2); the threefold torsional barrier can be readily deduced from the threefold axis of symmetry of ethane.

The molecule of butane (II) provides a more complex example. Despite the fact that this molecule is a triple rotor, we will consider only rotation about the central bond (ethane rotation). Besides the innumerable skewed forms, three eclipsed and three staggered conformations exist. As opposed to ethane, the three staggered conformations are not equivalent. In structure IId, the two methyl groups are trans,* and therefore, it is called the

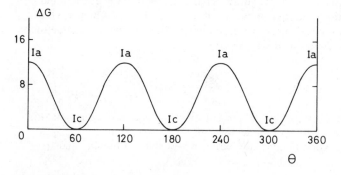

FIGURE 2 Conformational energy (in kJ/mol) of ethane as a function of torsion angle.

*Also designated as anti.

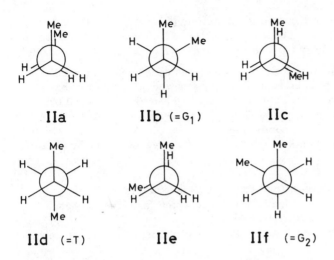

trans-conformer (T). The conformers IIb and IIf are <u>gauche</u> (G$_1$ and G$_2$); it is apparent that they are enantiomeric and therefore of identical energy. (The gauche-forms are less stable than the trans-form by about 3.3 kJ/mol or 0.8 kcal/mol in the liquid state.)

Regarding the eclipsed conformations of butane (transition states of interconversion), once again they are nonequivalent; maximal steric strain exists in conformation IIa, where the two methyl groups reach their closest approach. The enantiomeric conformations IIc and IIe, as compared to IIa, replace one Me/Me and one H/H interaction by two Me/H interactions; this is accompanied by an energy gain of about 13.8 kJ/mol (3.3 kcal/mol). These facts are summarized in Figure 3, which clearly

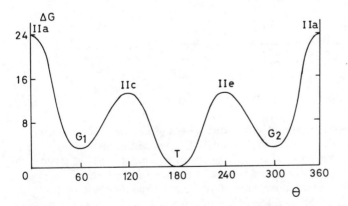

FIGURE 3 Conformational energy (in kJ/mol) of <u>n</u>-butane as a function of rotation about the central bond.

shows a global minimum (T), two local minima (G_1 and G_2), a transition state of higher energy (IIa; about 25 kJ/mol or 6 kcal/mol), and two transition states of lower energy (IIc and IIe; about 15.5 kJ/mol or 3.7 kcal/mol).

The designation of rotational conformers as used above is not fully satisfactory, and tends to be superseded by a newer nomenclature based partly on the sequence rule. This nomenclature has been proposed by Klyne and Prelog [7] and starts by defining the torsion angle A-X-Y-B (usually designated theta, θ, sometimes also tau, τ). A and B are selected from the sets of substituents carried by X and Y, respectively, by the following criteria:

If all substituents are different, the sequence rule is applied.
If two substituents are identical, the one which is unique is chosen
independently of the sequence rule.
If all substituents are identical, the one providing the smallest torsion
angle is chosen.

When viewing A-X-Y-B along X-Y as shown in structure III, the torsion angle is defined by the angle formed by the segments XA and YB. The sign of the torsion angle is defined by the rotation which brings A to overlap with B; the angle is positive for a clockwise rotation, and negative for a counterclockwise rotation [(P)- and (M)-helicity, respectively]. This definition of the torsion angle permits a clear description of the A-X-Y-B angle, as opposed to the dihedral angle, Φ (angle between the planes A-X-Y and X-Y-B), where the sign of rotation is not apparent (see Chapter 3).

The nomenclature of torsional isomers, based on the torsion angle, divides the circle (360° rotation) into several fields as shown in structure IV; the resultant designations are listed in Table 1.

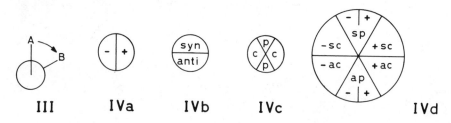

III   IVa   IVb   IVc   IVd

The through-space interactions influencing the stability of rotamers have been considered at length in Section 3.3. Due to the low-energy barrier and the minute energy differences between rotamers, minor structural alterations may result in considerable conformational changes. For example, the preferred conformation of several ethane derivatives is gauche, as opposed to the usual preference for the trans-rotamer. Because the energy terms involved in conformational control and in solvation are of

TABLE 1   Description of Steric Relationships Across Single Bonds[a]

Torsion angle θ	Designation
$0° \pm 30°$	$\pm$ Synperiplanar ($\pm$sp)
$+30°$ to $+90°$	$+$ Synclinal ($+$sc)
$+90°$ to $+150°$	$+$ Anticlinal ($+$ac)
$+150°$ to $+180°$	$+$ Antiperiplanar ($+$ap)
$-30°$ to $-90°$	$-$ Synclinal ($-$sc)
$-90°$ to $-150°$	$-$ Anticlinal ($-$ac)
$-150°$ to $-180°$	$-$ Antiperiplanar ($-$ap)

[a]See drawings IV.   From Klyne and Prelog [7].

comparable amplitude, the conformational behavior of a given molecule will vary greatly with its environment.

To illustrate the above discussion quantitatively, a few examples are given in Tables 2 and 3. The barriers of rotation reported in Table 2 concern compounds with a potential function having threefold symmetry. The influence of steric strain is apparent when comparing monosubstituted ethanes and 2,3,3-trimethylbutanes. A striking value is the high barrier of CF_3-CCl_3, which probably originates from both van der Waals and electrostatic repulsive interactions between the halogen atoms.

A recently reported and interesting example is that of 2,3-dimethyl-butane (V). The <u>anti</u>-conformer (Va) isomerizes to the <u>gauche</u>-forms (Vb

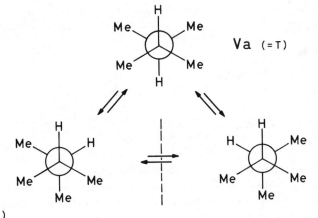

TABLE 2   Barrier to Rotation in Model Molecules

	Barrier	
Compound	kJ/mol	kcal/mol
CH_3 - CH_2 X^a		
X = H	12.1	2.88
X = F	13.9	3.33
X = Cl	15.4	3.68
X = Br	15.4	3.68
X = I	13.4 ± 2.1	3.2 ± 0.5
CH_3 - CF_3 ^{b}	12.7	3.04
CH_3 - CCl_3	12.2	2.91
DF_3 - CCl_3	25.1	6.0
$(CH_3)_3C$ - $C(CH_3)_2X^c$		
X = H	29.2	6.97
X = F	33.7	8.04
X = Cl	43.7	10.43
X = Br	44.9	10.73
X = I	46.6	11.14

^{a}From Lowe [8].

^{b}From Millen [9].

^{c}From Anderson et al. [10].

and Vc) with a barrier of 18.0 kJ/mol (4.3 kcal/mol) involving two Me/H and one Me/Me eclipsed interactions; the direct G_1 to G_2 conversion must overcome a barrier of about 33 kJ/mol (8 kcal/mol) [11].

With regard to the relative energy of staggered rotamers (Table 3), the predominance of antiperiplanar forms is verified in most cases. For butane (II), the relief of the gauche Me/Me interaction in structures IIb and IIf is often accepted as the origin of the trans predominance. However,

TABLE 3 Stability of Rotamers

Compound	X	Y	$E_{gauche} - E_{trans}$[a]			
			Gas		Liquid	
			kJ/mol	kcal/mol	kJ/mol	kcal/mol
$XCH_2 - CH_2Y$[b]	CH_3	CH_3[c]	4.0	0.96[d]	3.2	0.76
	Cl	Cl	4.6	1.1	0	0
	Br	Br	6.3	1.5	2.9	0.7
	Br	Cl	5.9	1.4	2.1	0.5
	CH_3	Cl	-0.2	-0.05	0.2	0.05
	CH_3	Br	-0.63	-0.15	—	—
	Cl	OH[e]	-4.0	-0.95	-4.0	-0.95
$X_2CH - CHY_2$[b]	Cl	Cl	0	0	-4.6	-1.1
	Br	Br	—	—	-4.0	-0.95

Compound	X	Solution in CF_2Cl_2	
		kJ/mol	kcal/mol
$(CH_3)_2\ C - C(CH_3)_2$[f]	H	1.2	0.29
$\quad\ \ \mid\quad\ \mid$	F	-1.9	-0.45
$\quad\ \ H\quad X$	Cl	0.42	0.10
	Br	0.88	0.21
	I	2.3	0.56

[a]Positive values which signify <u>trans</u> are preferred.

[b]From Millen [9].

[c]See structure II.

[d]From Verma et al. [11a].

[e]See structure VI.

[f] From Anderson et al. [10].

this is far from being the complete story. Gauche H/H interactions are also operative; they have been postulated [12] to be the main destabilizing factor, but it now established [13,14] that the overall contribution of the gauche hydrogen effect to the total gauche/trans conformational energy cannot overbalance other contributions (see Section 3.3). In addition to these nonbonded steric repulsions, "intrinsic" ("nonsteric") contributions to gauche destabilization exist and cannot be neglected; they comprise hyperconjugative and other through-bond interactions which make up one- and twofold barrier components [15].

In the above context, it is worthwhile mentioning that the trans- and gauche-conformers of 2,3-dimethylbutane (V) have practically identical energies, i.e., the $T:G_1:G_2$ ratios are 1:1:1 (T:G ratio = 1:2) [11,16].

Formation of an intramolecular hydrogen bond as seen in the enantiomeric gauche-forms VIb and VIc explains the conformational behavior of 2-chloroethanol (Table 3); this effect is even more pronounced in 2-fluoroethanol where the gauche-form predominates by about 8 kJ/mol

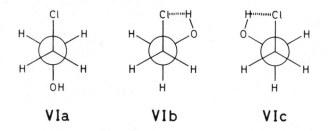

VIa   **VIb**   **VIc**

(2 kcal/mol) [17]. Any discussion of rotamer stability based on considerations of pure staggered conformers is nothing more than an approximation. Indeed, dihedral angles of 60° are assumed, but it is known that marked deviations do exist. Also, bond length and bond angle distortions tend to distribute globally the strain formally generated by the gauche interactions. For a realistic assessment of rotamer stability one must, therefore, consider all structural components. This has been nicely illustrated in the case of butane [12], although the numerical values reported have been refined.

After having discussed rotation about the sp^3-sp^3 carbon-carbon bond, let us now turn our attention toward sp^3 carbon-heteroatom bonds. The rotation barriers are usually low, unless strong nonbonded interactions between the substituents become predominant. Factors governing the values reported in Table 4 include the carbon-heteroatom bond length, the number of H/H interactions, and partially understood electronic factors

TABLE 4 Barrier of Rotation[a]

Compound	Barrier	
	kJ/mol	kcal/mol
$CH_3 - CH_3$	12.1	2.88
$CH_3 - SiH_3$	7.12	1.70
$CH_3 - NH_2$	8.13	1.94
$CH_3 - PH_2$	8.21	1.96
$CH_3 - OH$	4.48	1.07
$CH_3 - SH$	5.32	1.27
$CH_3 - NO_2$	0.025	0.006

[a]From Gordon and Ford [1] and Millen [9].

(e.g., electronic density of hydrogen atoms and exchange interactions between X-H and C-H orbitals).

The case of nitromethane is especially interesting. The planar nitro group with twofold symmetry, and the methyl group with threefold symmetry, result in a sixfold barrier of rotation whose energy is exceedingly low (Table 4).

Ethanol provides a more complex case than the molecules reported in Table 4. Three conformers exist, a trans-form (VIIa) and the two enantiomeric gauche-forms (VIIb and VIIc). Quantum calculations indicate the trans-conformer to have a relative stability of 2.64 kJ/mol (0.63 kcal/mol); the trans-gauche barrier is 5.61 kJ/mol (1.34 kcal/mol), comparable to the methanol barrier (Table 4), while the gauche-gauche barrier is

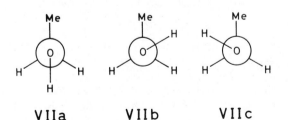

VIIa   VIIb   VIIc

8.3 kJ/mol (2.06 kcal/mol) [18]. Torsional constants for molecular fragments of alcohols and ethers have been calculated [19].

Because the inversion barriers of oxygen and nitrogen (see Section 5.2) are somewhat above the rotation barriers reported in Table 4, the introduction of bulky substituents may result in rotation being obscured by inversion. In the amino derivative VIII, dynamic nuclear magnetic resonance (see Section 12.6) has allowed observation of both processes. The rotation about the C-N central bond was found to have a barrier of 23.5 kJ/mol (5.6 kcal/mol), while inversion at the nitrogen atom requires 31.4 kJ/mol (7.5 kcal/mol) [20]. In the case of tert-butylamines inversion (IXa - IXb interconversion), but not C-N bond rotation (IXa - IXc interconversion), is observed [21].

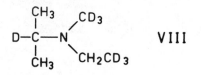 VIII

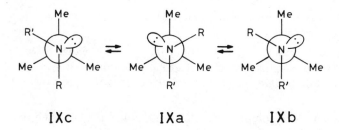

IXc   IXa   IXb

The rotation about heteroatom-heteroatom bonds is influenced by destabilizing interactions between the electron lone pairs [22]. Such interactions appear to play a marked role in controlling the conformational behavior of hydrogen peroxide [23], hydrazines [24], and hydroxylamines. An interesting example of this group of compounds is provided by the simple disulfide CH_3-S-S-CH_3 (X). The lowest energy rotamers have a dihedral

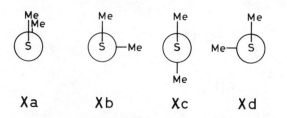

Xa   Xb   Xc   Xd

angle of 85-90° (i.e., the enantiomeric rotamers Xb and Xd); these are separated by a <u>trans</u> barrier (Xc) of 9.2 kJ/mol (2.2 kcal/mol) and an eclipsed barrier (Xa) of 29 kJ/mol (7.0 kcal/mol) [25].

9.3 Multiple Rotors

When several rotors are present in a molecule, they show some degree of concertedness in their rotation. Taking <u>n</u>-propane as an example, its preferred conformation is the fully staggered one (XIa). The rotation barrier of propane is reported in Table 5 together with that of analogous compounds; it is, however, necessary to note that this barrier, which is the one observed experimentally, involves rotation of <u>one</u> methyl group alone, the other group being in its preferred conformation (XIb). Thus, the data in Table 5 do not refer to the higher energy transition state (e.g., XIc), having the two methyl groups in eclipsed conformations (see also the discussion on dimethylsulfone in the next section).

TABLE 5 Rotation Barriers for Multiple Rotors[a]

Compound	X	Barrier	
		kJ/mol	kcal/mol
		(Two threefold barriers)	
$CH_3 - X - CH_3$	CH_2	13.8	3.3
	SiH_2	6.9	1.65
	NH	13.8	3.3
	PH	9.2	2.2
	O	10.5	2.5
	S	8.8	2.1
	Se	6.3	1.5
		(Three threefold barriers)	
$(CH_3)_3 X$	CH	16.3	3.9
	N	18.4	4.4
	P	10.9	2.6

[a]From Gordon and Ford [1], Millen [9], and Lambert and Featherman [26].

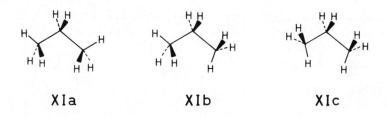

XIa **XIb** **XIc**

   A comparison of the data in Tables 4 and 5 shows a regular increase in rotation barrier with increasing substitution for all homologous systems (one, two, three rotors; ethane, propane, isobutane; mono-, di, tri- methylamine; methanol, dimethyl ether; methanethiol, dimethyl sulfide). This phenomenon is accounted for by repulsive interactions between the hydrogen atoms of the different methyl groups [9]. Recently, the changes of the rotation barrier when going from a single rotor to the corresponding double rotor have been explained by a stabilization involving bonding between methyl groups, if the central atom is a π donor (e.g., dimethyl ether), and a destabilization if the central atom is a π acceptor [27].

   The preferred conformation of isobutane, like that of n-propane, is the fully staggered one (XII). In the case of n-pentane (XIII), the trans-trans-conformer (designated T.T) is somewhat favored over the trans-gauche-conformer (T.G). In general, a methylenic chain exhibits some tendency to exist as the fully extended conformer (T.T.T...). However, it may not be predominant because its statistical weight is small as compared to the sum of all other conformers.

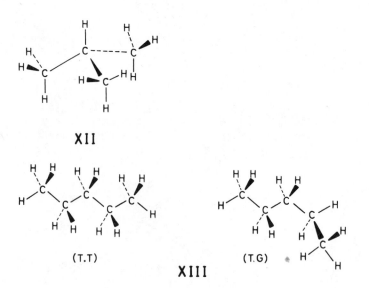

XII

(T.T)   (T.G)

XIII

In the case of substituted polymethylene compounds, the number of conformers markedly increases. Taking 1,3-dichloropropane (XIV) as an example, the conformers are: (T.T) (statistical weight = 1x); (T.G) (4x); (G.G) (2x); and, (G.G') (2x). As opposed to hydrocarbon chains (e.g., n-pentane), the (G.G)-conformer predominates over the (T.T)-form. On the other hand, the (G.G')-conformer is improbable on steric grounds [9]. Several additional examples can be found in a recent review article dealing with alcohols and amines [18].

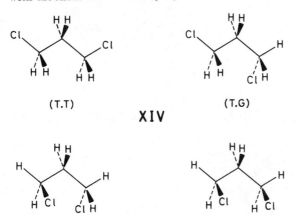

(T.T)   (T.G)

XIV

(G.G)   (G.G')

9.4 Conjugated Systems

The presence of sp^2-hybridized carbon atoms considerably influences the conformational behavior of molecules about single bonds. We shall consider successively sp^2-sp^3 carbon-carbon bonds, sp^2-sp^2 carbon-carbon bonds, and sp^2 carbon-heteroatom bonds.

Rotational isomerism about sp^2-sp^3 hybridized carbon-carbon single bonds can be discussed in terms of rotamers XVa-d [28]. A much-studied class of compounds are the aliphatic aldehydes (X = O; Y = H). Acetaldehyde has a threefold barrier to rotation which is small (about 4.6 kJ/mol

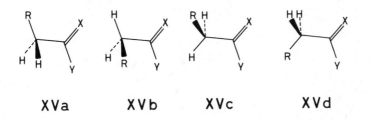

XVa   **XVb**   **XVc**   **XVd**

or 1.1 kcal/mol), the preferred conformation being a slightly eclipsed form (by 9°) (XVI). Similarly, propionaldehyde exists in eclipsed forms (XVIIa and XVIIc), with form XVIIa being favored over form XVIIc by about 3.8 kJ/mol (0.9 kcal/mol). Further, it has been shown that XVIIc is, in fact, not fully eclipsed, but distorted by about 11° [29]. The curve of conformational energy versus torsion angle [30] is shown in Figure 4, with the O/Me and O/H eclipsed conformers being global and local minima, respectively. The H/H eclipsed conformers (two enantiomeric forms, XVIIb and XVIIf) are the higher energy transition states (about 8.8 kJ/mol or 2.1 kcal/mol), and the H/Me eclipsed conformer (XVIId) is the lower energy transition state (6.3 kJ/mol or 1.5 kcal/mol) [16].

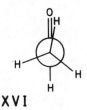

XVI

In general, it appears that aldehydes exist mainly as rotamers XVa and XVb, with XVa (O/R eclipsed form) predominating. Bulky substituents R tend to shift the balance between the two forms, and the influence of the solvent is dependent on the nature of the substituent, R [28].

Carboxylic acids and esters (structure XV: X = O; Y = OH and OR', respectively) are conformationally comparable to aldehydes. These

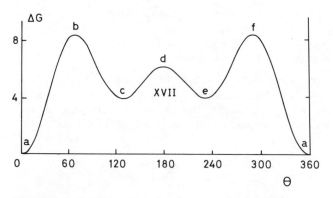

FIGURE 4 Conformational energy (in kJ/mol) of propionaldehyde as a function of rotation about the sp^2-sp^3 carbon-carbon bond. From Pickett and Scroggin [30].

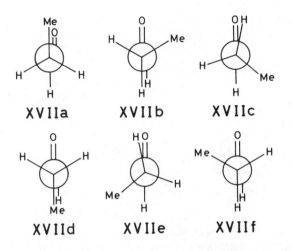

XVIIa   **XVIIb**   **XVIIc**

XVIId   **XVIIe**   **XVIIf**

molecules are, in fact, double rotors. Rotation about the sp^2-hybridized carbon-oxygen bond was shown in the case of formic acid (XVIII) and ethyl formate (XVIII: R' = ethyl) to yield two preferred conformations, namely the <u>trans</u>- (XVIIIa) and the <u>cis</u>-form (XVIIIb). In the case of formic acid and probably also ethyl formate, the <u>cis</u>-form is the more stable by about 8 kJ/mol[9]. In acetic acid, the rotation about the C-C bond has a barrier of about 2 kJ/mol, the preferred rotamer being the H/O eclipsed form. Comparable results have been obtained for methyl acetate. Therefore, the predominant conformer of these two compounds should be structure XIX, in which the atoms HCCOOH(C) are coplanar [9,28].

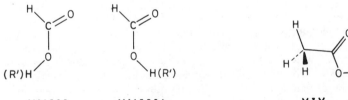

XVIIIa   **XVIIIb**   **XIX**

In agreement with the behavior of aldehydes, <u>aliphatic ketones</u> predominantly exist as eclipsed rotamers. Acetone displays the preferred conformation depicted in structure XX, with a rotation barrier of about 3.3 kJ/mol (0.8 kcal/mol), a relatively small value [9,28]. Similarly, the most stable rotamer of diethyl ketone is structure XXI.

Olefinic molecules containing the propene moiety (i.e., structure XV: X = CH_2) show close similarities with the carbonyl compounds discussed above. The preferred conformation of propene itself shows a hydrogen

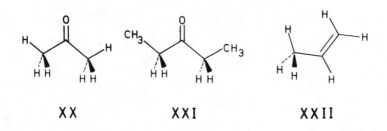

XX　　　　　XXI　　　　　XXII

atom eclipsing the double bond (XXII); the rotation of the methyl group has
a barrier of 8.4 kJ/mol (2.0 kcal/mol) [9]. Mono-3-substituted propenes
have two low-energy conformations differing by the position of the
3-substituent; rotamer XXIIIb is favored for most derivatives (e.g.,
1-butene: R = CH$_3$; about 1.75 kJ/mol or 0.4 kcal/mol) while 3-fluoro-
propene, 3-methoxypropene, and a few others show some preference for
rotamer XXIIIa. Usually, the trends observed with 3-substituted propenes
and acetaldehyde are quite similar (influence of the size of R, and of the
solvent, etc.).

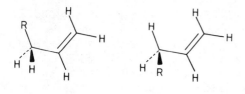

XXIIIa　　　　　XXIIIb

   The phenyl derivatives are structurally close to the above molecules,
as exemplified by toluene (XXIV), a compound which again shows eclipsing
between a hydrogen atom and the π system. The rotation barrier of toluene
is extremely low, a few joules per mole. Indeed, toluene is like nitromethane
(Table 4), a compound having a sixfold barrier, and the latter is always
found to be extremely small. In such molecules an eclipsed conformation
is to be found after every 60° rotation, and no real relief of conformational strain is obtained after a rotation of only 30°; the barrier is
minute simply because no low-energy conformers exist.

XXIV

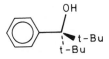

XXV

$$X-\bigcirc-\underset{\underset{CH_3}{|}}{\overset{\overset{Y}{|}}{C}}-\underset{\underset{CH_3}{|}}{\overset{\overset{CH_3}{|}}{C}}-CH_3 \qquad \textbf{XXVI}$$

When bulky substituents are introduced in toluene in an unsymmetrical way, the rotation barrier becomes quite large; for example, it is close to 80 kJ/mol for compound XXV [31]. An interesting example has been studied recently, namely, derivatives of 1,2,2-trimethyl-1-phenylpropane (XXVI); for Y = H, and X = NO_2 or OCH_3, the barriers of rotation were found to be 37.7 and 36.9 kJ/mol (9.0 and 8.8 kcal/mol), respectively; for Y = Cl, and X = NO_2 or OCH_3, the barriers are 32.7 and 31.0 kJ/mol (7.8 and 7.4 kcal/mol), respectively [31]. This example points to two distinct effects. First, electronic effects of the para-substituent play a minor, but clear role in influencing the rotation about the sp^2-sp^3 bond. Second, the lower barrier of the 1-chloro derivatives results from the increased energy of the preferred conformation having Y in the plane of the ring; this destabilization (about 4 kJ/mol) of the preferred conformation is due to a repulsive interaction between the chlorine atom and an ortho-hydrogen [31].

The conformational behavior about sp^2-sp^3 bonds as compared to sp^3-sp^3 bonds has received due attention from theoreticians and experimentalists. Quantum calculations have shown [32,33] that the main factors controlling the conformation of sp^2-sp^3 systems are one-electron attractive interactions, whereas in sp^3-sp^3 systems they are two-electron repulsive interactions. It was found that the energy of the one-electron attractive interactions decreases with decreasing electron density in the methyl group. In both propene and acetaldehyde, interaction of the electron density of the methyl C-H bonds with the π-electron density occurs, and is called hyperconjugation. As the methyl group rotates, its σ- and π-electron loss to the double bond varies, the preferred conformation showing minimal loss, i.e., maximal electron density and therefore maximal one-electron attractive interactions [32].

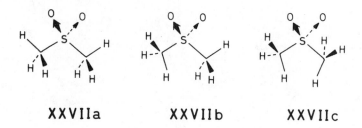

XXVIIa   **XXVIIb**   **XXVIIc**

Another case involving hyperconjugation is represented by dimethyl-sulfone (XXVII). Quantum calculations confirmed experimental works and showed the preferred conformation to be structure XXVIIa; the barrier of rotation of one methyl group (with XXVIIb as transition state) was found to require 15.3 kJ/mol (3.65 kcal/mol). Simultaneous rotation of the two methyl groups has been demonstrated to be a higher energy process, with structure XXVIIc being 35.0 kJ/mol (8.35 kcal/mol) above the global minimum. It was also demonstrated that a hyperconjugative S/H interaction accounts for about 70% of the conformational energy, while the remainder is mainly accounted for by O/H and H/H (methyl/methyl) interactions [34].

The factors controlling the conformational behavior about sp^2-sp^2 hybridized carbon-carbon single bonds differ from those involved in rotational isomerism about sp^2-sp^3 bonds. In both cases, however, they are of an electronic nature. It is well known that the two double bonds experience conjugation across the single bond, and that this phenomenon is most pronounced when the system is planar (see Section 3.3).

In general, rotation about sp^2-sp^2 carbon-carbon single bonds will result in planar-preferred conformations, and out-of-plane (conceivably perpendicular) transition states. In many systems, two planar conformations exist, namely, those having the two double bonds trans and cis relatively to the single bond (e.g., XXVIIIa and XXVIIIb, respectively). The two isomers are referred to as s-trans (or antiperiplanar) and s-cis (or syncoplanar), the single bond (s) having in fact a partial π character (see Section 3.3). In simple compounds, the s-trans-rotamer is favored over its s-cis-isomer for mere steric reasons. Substitution at some of the carbon atoms may render the trans-conformer more crowded and reverse the rule of trans preference.

The conformational parameters of a few selected compounds are represented in Table 6. These compounds are conjugated dienes, α,β-unsaturated carbonyls, and aromatic aldehydes. The increased stability of the s-trans-rotamer of methacrolein (XXXa) as compared to acrolein (XXIX) is believed to be due to an attractive electrostatic interaction between the polarizable α-methyl group and the carbonyl oxygen [17]. The conformational energy of variously methylated α,β-unsaturated aldehydes and ketones has been calculated and discussed [38].

TABLE 6 Conformation Energy About sp^2-sp^2 Single Bonds[a]

Compound	Structure	$E_{s-cis} - E_{s-trans}$ $(E_b - E_a)$		Barrier	
		kJ/mol	kcal/mol	kJ/mol	kcal/mol
Butadiene	XXVIII	10.5	2.5	30.0	7.16
Acrolein	XXIX	8.63	2.06	20.8	4.96
Methacrolein	XXX	13.0	3.1	22.2	5.3
Benzaldehyde	XXXI	0	0	20.5	4.9
Furfuraldehyde	XXXII	4.2	1.0	46.1	11.0

[a]From Wyn-Jones and Pethrick [17], Laurie [35], Abraham and Siverns [36], and Carreira [37].

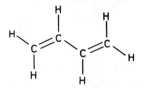

XXVIIIa

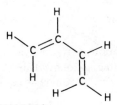

XXVIIIb

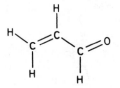

XXIXa

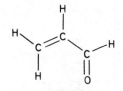

XXIXb

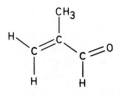

XXXa

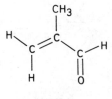

XXXb

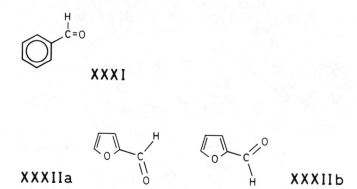

XXXI

XXXIIa   **XXXIIb**

Benzaldehyde (XXXI) shows degeneracy of s-trans- and s-cis-forms, a phenomenon, of course, not occurring in the case of mono-ortho-substitution [39]. Furfuraldehyde (XXXII), as the pure liquid (value reported in Table 6), and in polar solvents, shows preference for the trans-form XXXIIa (often called cis with respect to the two oxygens). In the absence of dominant solvation factors, the cis-form XXXIIb is preferred because of charge repulsion between the two oxygen atoms [36]; only in polar media is the charge repulsion sufficiently diminished to allow the "normal" trans preference to emerge.

Biphenyls offer a case of particular interest, which has been the subject of many studies and discussions. Biphenyl itself (XXXIIIa) experiences two opposing forces, namely, resonance stabilization favoring a planar

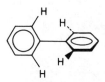

XXXIIIa   **XXXIIIb**

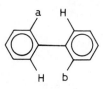   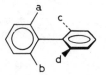

XXXIIIc   **XXXIIId**

TABLE 7 Barrier of Rotation (Racemization) for Biphenyls

	Barrier	
Compound	kJ/mol	kcal/mol
2,2'-Disubstituted biphenyls[a] (XXXIIIb, c)		
a = b = OCH$_3$	57.4	13.7
a = b = CH$_3$	72.9	17.4
a = b = OCOCH$_3$	77.5	18.5
a = b = CH$_2$OCOCH$_3$	84.6	20.2
a = b = CH(CH$_3$)$_2$	>109.0	>26.1
2,2',6,6'-Tetrasubstituted biphenyls and analogs[b] (XXXIIId)		
a = H; b = OEt; c = NO$_2$; d = COOH	83.8	20.0
a = c = OCH$_3$; b = d = COOCH$_3$	105.0	25.0
a = c = OCH$_3$; b = d = CONH$_2$	116.0	27.8
a = c = NO$_2$; b = d = CH$_2$— N —CH$_2$ (bridge) CH$_3$	126.0	30.0

[a]From Kessler [41].

[b]From Hall and Harris [42].

conformation, and steric interactions between the ortho-hydrogens which favor clinal conformations. Biphenyl has been found to be planar only in the crystal; in solution, the two rings are estimated to be inclined by approximately 20°. The interplanar angle will increase with the increasing bulk of ortho-substituents [40].

The rotation in 2,2'-substituted biphenyls (XXXIIIb, c) involves two diastereoisomeric transition states, both of which are planar: the cisoid-form (XXXIIIb) and the transoid-form (XXXIIIc), the latter showing less severe overcrowding. Table 7 shows that the energy barriers are relatively high, and that some of them resort to configurational rather than conformational isomerism. This is even more evident with 2,2',6,6'-tetrasubstituted derivatives (Table 7). If the minimum conditions for a chiral axis are met (e.g., a ≠ b and c ≠ d, see structure XXXIIId), and if the barrier allows fair stability, suitable biphenyls do exist as resolvable enantiomers (see Section 7.1).

Stereoisomerism due to restricted rotation about single bonds is called atropisomerism, and separable single bond torsional isomers are atropisomers. By our definition, atropisomerism resorts to configurational isomerism at room temperature. Atropisomers may be enantiomers (e.g., suitable biphenyls) or diastereoisomers (e.g., several substituted para-terphenyls which will not be discussed in these pages).

Rotation about sp^2 carbon-heteroatom single bonds is usually a higher energy process compared to rotation about sp^3 carbon-heteroatom single bonds (see Section 9.2) due to conjugation and to the resulting partial double bond character of the pivot bond. If we consider first sp^2 carbon-oxygen single bonds, interesting compounds encountered include phenols, aromatic ethers, and all classes of carboxylate esters. Table 8 lists a few values for phenol and aromatic ethers.

TABLE 8 Barrier of Rotation About sp^2 Carbon-Oxygen Single Bonds in Aromatic Compounds[a]

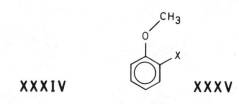

XXXIV   XXXV

| Compound | Structure | Barrier | |
		kJ/mol	kcal/mol
Phenol	XXXIV	13.0	3.1
Anisole	XXXV; X = H	25.0	6.0
ortho-Fluoroanisole	XXXV; X = F	26.2	6.25
ortho-Chloroanisole	XXXV; X = Cl	37.3	8.9
ortho-Bromoanisole	XXXV; X = Br	38.1	9.1

[a]From Millen [9], and Wyn-Jones and Pethrick [17].

Carboxylate esters exist as two preferred rotamers, namely the two planar s-cis- (XXXVIa) and s-trans-forms (XXXVIb). While the barrier of rotation remains close to 34 kJ/mol (about 8 kcal/mol) for simple aliphatic esters, the energy difference between the s-cis- and s-trans-rotamers increases with the total number of carbon atoms in both R_1 and R_2 (XXXVI). The s-cis-conformer is favored by about 8 kJ/mol in formic acid esters, and by a larger energy difference in higher homologs [17]. These results imply that increasing the size of R_1 and R_2 destabilizes the s-trans-form, but does not influence significantly the s-cis-form and the clinal transition state [17].

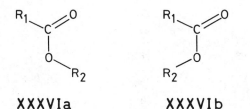

XXXVIa   **XXXVIb**

Torsional isomerism about sp^2 carbon-nitrogen single bonds exists in compounds such as amides, imides, anilines, anilides, and others. Amides, in particular, have been the object of numerous studies. Maximal conjugation markedly favors a planar conformation; this phenomenon is often represented by the canonical structures XXXVIIa and XXXVIIb.

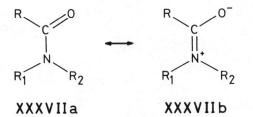

XXXVIIa   **XXXVIIb**

The importance of the partial double bond character is well-illustrated by the relatively high barriers of rotation of amides. It can be seen in Table 9 that the height of the barrier seemingly decreases with the increasing steric bulk of the substituents. This is attributed to a relative destabilization of the ground state, but does not affect the transition state

TABLE 9 Conformational Behavior of Simple Amides[a]

Amide XXXVII			Rotation barrier	
			kJ/mol	kcal/mol
R	R_1	R_2		
H	H	H	74.6	17.8
H	CH_3	CH_3	88.0	21.0
CH_3	H	H	70.8	16.9
CH_3	CH_3	CH_3	75.8	18.1
Ethyl	CH_3	CH_3	72.1	17.2
Isopropyl	CH_3	CH_3	67.9	16.2
Phenyl	CH_3	CH_3	67.0	16.0
CH_3	Ethyl	Ethyl	70.8	16.9
CH_3	n-Propyl	n-Propyl	71.2	17.0
CH_3	Isopropyl	Isopropyl	65.8	15.7

Amide XXXVII			Percent of (Z)-isomer
R	R_1	R_2	
H	CH_3	H	92
H	Ethyl	H	88
H	Benzyl	H	90
CH_3	CH_3	H	97 ~ 100
Ethyl	CH_3	H	~ 100
Ethyl	Ethyl	H	~ 100
H	Ethyl	CH_3	40
CH_3	Ethyl	CH_3	51
H	t-Butyl	CH_3	11
CH_3	t-Butyl	CH_3	53

[a]From Kessler [41], Steward and Siddall [43], and Noe and Raban [44].

itself. However, the possibility of superimposed electronic effects of inductive nature cannot be excluded, and, in some cases, appears to over-whelm the steric effects [44].

When R_1 and R_2 (XXXVII) are different substituents, two planar rotamers exist whose designation is not devoid of ambiguity. Thus, XXXVIII is usually designated <u>s-cis</u>, but some authors name it <u>trans</u> due to the relative positions of R and R_1. A better way to designate such rotamers is to apply the E and Z nomenclature (see Section 8.1) to the canonical structure XXXVIIb having the double bond between the carbonyl carbon and the amide nitrogen [44]; assuming $O > R$ and $R_1 > H$, structure XXXVIII is the (Z)-form.

XXXVIII

In the case of N-monosubstituted amides, the (Z)-isomer predominates by more than 4 kJ/mol over the (E)-form (Table 9). Steric repulsion be-tween R and R_1, as well as attractive electrostatic interactions between R_1 and the carbonyl oxygen seemingly account for this preference. In the case of N,N-disubstituted amides with dissimilar substituents of compara-ble steric bulk, marked isomeric predominance is lost (Table 9).

References

1. A. J. Gordon and R. A. Ford, The Chemist's Companion, Wiley, New York, 1972, pp. 115-124, 156-167.

2. O. Hassel, Tidssk. Kjemi Bergv. Metal., 3:32 (1943).

3. D. H. R. Barton, Experientia, 6:316 (1950).

4. Topics in Stereochemistry, vol. 6 (edited by N. L. Allinger and E. L. Eliel), Wiley, New York, 1971, pp. 1-17.

5. E. L. Eliel, N. L. Allinger, S. J. Angyal, and G. A. Morrison, Conformational Analysis, Wiley, New York, 1965.

6. M. Hanack, Conformation Theory, Academic Press, New York, 1965.

7. W. Klyne and V. Prelog, Experientia, 16:521 (1960).

8.   J. P. Lowe, Prog. Phys. Org. Chem., 6:1 (1968).

9.   D. J. Millen, Restricted rotation about single bonds, in Progress in Stereochemistry, vol. 3 (edited by P. B. D. de la Mare and W. Klyne), Butterworths, London, 1962, pp. 138-168.

10. J. E. Anderson, C. W. Doecke, and H. Pearson, J. Chem. Soc. P.T.II., 336 (1976).

11. L. Lunazzi, D. Macciantelli, F. Bernardi, and K. U. Ingold, J. Am. Chem. Soc., 99:4573 (1977).

11a. A. L. Verma, W. F. Murphy, and H. J. Bernstein, J. Chem. Phys., 60:1540 (1974).

12. D. H. Wertz and N. L. Allinger, Tetrahedron, 30:1579 (1974).

13. S. Fitzwater and L. S. Bartell, J. Am. Chem. Soc., 98:5107 (1976).

14. E. Ōsawa, J. B. Collins, and P. v. R. Schleyer, Tetrahedron, 33:2667 (1977).

15. L. S. Bartell, J. Am. Chem. Soc., 99:3279 (1977).

16. N. L. Allinger, Calculation of molecular structure and energy by force-field methods, in Adv. Phys. Org. Chem., vol. 13 (edited by V. Gold and D. Bethell), Academic Press, London, 1976, pp. 1-83.

17. E. Wyn-Jones and R. A. Pethrick, The use of ultrasonic absorption and vibrational spectroscopy to determine the energies associated with conformational changes, in Topics in Stereochemistry, vol. 5 (edited by E. L. Eliel and N. L. Allinger, Wiley, New York, 1970, pp. 205-274.

18. D. R. Truax and H. Wieser, Chem. Soc. Rev., 5:411 (1976).

19. N. L. Allinger and D. Y. Chung, J. Am. Chem. Soc., 98:6798 (1976).

20. C. H. Bushweller, C. Y. Wang, J. Reny, and M. Z. Lourandos, J. Am. Chem. Soc., 99:3938 (1977).

21. W. R. Jackson and W. B. Jennings, Tetrahedron Lett., 1837 (1974).

22. J. Hine, Structural Effects on Equilibria in Organic Chemistry, Wiley, New York, 1975, pp. 103-129.

23. P. G. Butron and B. R. Markey, Austral. J. Chem., 30:231 (1977).

24. M. J. S. Dewar and W. B. Jennings, J. Am. Chem. Soc., 95:1562 (1973).

25. D. B. Boyd, J. Am. Chem. Soc., 94:8799 (1972).

26. J. B. Lambert and S. I. Featherman, Chem. Rev., 75:611 (1975).

27. D. Cremer, J. S. Binkley, J. A. Pople, and W. J. Hehre, J. Am. Chem. Soc., 96:6900 (1976).

28. G. J. Karabatsos and D. J. Fenoglio, Rotational isomerism about sp^2-sp^3 carbon-carbon single bonds, in Topics in Stereochemistry, vol. 5 (edited by E. L. Eliel and N. L. Allinger, Wiley, New York, 1970, pp. 167-203.

29. S. S. Butcher and E. B. Wilson, Jr., J. Chem. Phys., 40:1671 (1964).

30. H. M. Pickett and D. G. Scroggin, J. Chem. Phys., 61:3954 (1974).

31. J. E. Anderson, H. Pearson, and D. I. Rawson, J. Chem. Soc. Chem. Comm., 95 (1973).

32. A. Liberles, B. O'Leary, J. E. Eilers, and D. R. Whitman, J. Am. Chem. Soc., 94:6894 (1972).

33. J. E. Eilers and A. Liberles, J. Am. Chem. Soc., 97:4183 (1975).

34. G. Robinet, F. Crasnier, J. F. Labarre, and C. Leibovici, Theor. Chim. Acta, 25:259 (1972).

35. V. W. Laurie, Acc. Chem. Res., 3:331 (1970).

36. R. J. Abraham and T. M. Siverns, Tetrahedron, 28:3015 (1972).

37. L. A. Carreira, J. Chem. Phys., 62:3851 (1975).

38. T. Liljefors and N. L. Allinger, J. Am. Chem. Soc., 98:2745 (1976).

39. L. Lunazzi, A. Ticca, D. Macciantelli, and G. Spunta, J. Chem. Soc. P.T.II, 1121 (1976).

40. D. M. Hall, The stereochemistry of 2,2'-bridged biphenyls, in Progress in Stereochemistry, vol. 4 (edited by B. J. Aylett and M. M. Harris), Butterworths, London, 1969, pp. 1-42.

41. H. Kessler, Angew. Chem. Int. Ed., 9:219 (1970).

42. D. M. Hall and M. M. Harris, J. Chem. Soc., 490 (1960).

43. W. E. Steward and T. H. Siddall, III, Chem. Rev., 70:517 (1970).

44. E. A. Noe and M. Raban, J. Am. Chem. Soc., 97:5811 (1975).

10

The Stereochemistry of Cyclic Systems

Cyclic systems show similarities with acyclic systems in many of their stereochemical aspects. However, distinct stereochemical charac- teristics exist which justify separate treatment. Also, the concepts of conformational isomerism, diastereoisomerism, and enantiomerism are often inseparable in such treatments.

10.1 Nonsubstituted Carbocycles

The cyclization of a chain generates a certain amount of strain; this can be determined experimentally by measuring the heat of combustion per CH_2 group and comparing it to the value for the acyclic analog (Table 1). The two main contributions to the cyclic strain are the Baeyer strain (angle strain, see Section 3.2) and the Pitzer strain (bond opposition strain of gauche and eclipsed conformations, see Section 9.2).

In small rings (3- and 4-membered), the Baeyer strain is especially significant, while it is less important or negligible in common rings (5- to 7-membered), in medium rings (8- to 11-membered), and in large rings (12- and higher membered). In all systems, the Pitzer strain is operative and will tend to be relieved by deviations from planarity of the carbon skeleton. In larger rings, transannular interactions also partially account for the relatively marked strain (Table 1).

It is trivial to state that cyclic molecules will tend to adopt the mini- mum energy conformation, that is, the one minimizing all strain contribu- tions. In this energy minimum, the remaining strain is optimally dis- tributed between the various contributions (bond length and bond angle deviations, Pitzer strain, and other nonbonded interactions). The cyclic systems to be discussed in this chapter usually exhibit several possible

TABLE 1 Strain in Carbocycles[a] (Calculated from Heat of Combustion Data, Assuming Cyclohexane to be Strainless)

Number of CH_2 in ring	Strain per CH_2 group		Strain for the molecule	
	kJ/mol	kcal/mol	kJ/mol	kcal/mol
3	38.6	9.2	116.0	27.6
4	27.2	6.5	109.0	26.0
5	5.0	1.2	25.1	6.0
6	0	0	0	0
7	3.8	0.9	26.4	6.3
8	5.0	1.2	40.2	9.6
9	5.9	1.4	52.8	12.6
10	5.4	1.3	54.5	13.0
12	2.9	0.7	35.2	8.4

[a]From Hanack [1] and Allinger et al. [2].

conformations; interconversions can occur by two distinct processes, namely, cycle inversion and pseudorotation. Cycle inversion involves a relatively high-energy transition state (some kilojoules per mole) occurring with modification of bond angles and of all other strains. Pseudorotation is a low-energy process which does not involve bond angle variations, but only changes in Pitzer strain and other nonbonded interactions. Conformers which can be transformed by pseudorotation are called flexible forms, while conformers which can only undergo inversion are called rigid forms.

Unsubstituted cyclopropane offers no stereochemical interest; it is planar by definition. The first homolog displaying conformational mobility is cyclobutane. The preferred conformation of cyclobutane is the puckered form I, which undergoes ring inversion to generate an identical

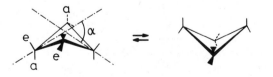

I

form. The inversion process can be compared to the wing motion of a butterfly. Because the puckered ground state is a relatively high-energy form (Table 1), the energy required to reach the planar transition state is minute (4.6 to 6.3 kJ/mol or 1.1 to 1.5 kcal/mol)[3,4]. This means the planar form will be detectably populated and may even be predominant for certain derivatives. The value of the puckering angle α is about 35° (I), corresponding to a torsion angle of 25° [5]. The conventional formula II is a convenient way to symbolize the preferred conformation of cyclobutane by indicating the values of the torsion angles. The exocyclic bonds in structure I belong to two different types, namely, the equatorial bonds (e) and the axial bonds (a).

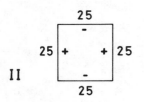

II

In cyclopentane, the bond angles have values close to the optimum; therefore, the strain in the molecule arises from bond opposition and is partly relieved by puckered conformations. Two flexible forms of cyclopentane have been differentiated. In the so-called envelope form (IIIa), one atom projects out of the plane of the four other atoms; this conformer has C_s symmetry (presence of a plane of symmetry σ). The other flexible form is the half-chair form (IIIb), in which three neighboring carbon atoms are coplanar, while the other two are above and below the plane, respectively, and equidistant from it. This conformer has C_2 symmetry [1,5,6].

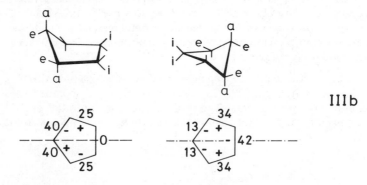

IIIa   IIIb

In cyclopentane, the envelope and half-chair forms interconvert through intermediate conformations with no symmetry. If in structure IIIa the out-of-plane atom (arbitrarily designated C-1 here) is pushed down together with one carbon next to it (C-2), a half-chair form is reached (C-1 above, C-2 below the plane). When the motion is continued, another envelope is reached, having C-1 in the plane and C-2 below it. The process then repeats with C-2 and C-3, and so on; 10 mutually indistinguishable envelope forms as well as 10 half-chairs interconvert by this process, which is known as pseudorotation. Indeed, it is not the molecule but the out-of-plane deformation that rotates, which is not unlike a wave on a water surface.

The pseudorotation circuit in cyclopentane is essentially of constant strain and, therefore, free; there are no energy minima and maxima [6]. But, the fully planar conformation is less stable by about 20 kJ/mol (about 5 kcal/mol) [6].

The two conformers IIIa and IIIb are extremes of higher symmetry in the pseudorotational circuit. In these forms, the following types of exocyclic bonds can be recognized: equatorial (e) and axial (a) bonds, as found in cyclohexane (see below), and the so-called isoclinal bonds (i). It must be noted, however, that according to the IUPAC Recommendations [7], the terms axial, equatorial, pseudoaxial, and pseudoequatorial may be applied to other than six-membered rings if, but only if, their interpretation is beyond dispute.

Among the alicycles, the most studied compound is without doubt cyclohexane. Its preferred conformation is a chair form, which was characterized in 1943 by Hassel [8]. An ideal chair form would have dihedral angles of 60°, and C-C-C bond angles of 109.5°. Because the normal $C-CH_2-C$ bond angle is 112.4°, the angle strain for the ideal chair form would be about 4 kJ/mol (about 1 kcal/mol). Also, a bond angle of 112.4° corresponds to dihedral angles of 52°, with a Pitzer strain of 2.9 to 3.3 kJ/mol (0.7 to 0.8 kcal/mol). The real cyclohexane chair balances these strains with bond angles of 111° and dihedral angles of 56°. In the hexagonal representation of cyclohexane (IVa), only the signs of the torsion angles are given; the angular values themselves can be omitted when they are 56°. Other torsion angle values are defined or given in brackets as deviations from the normal 56° value.

The chair conformation of cyclohexane shows equatorial and axial exocyclic bonds. This conformer is a rigid form which must undergo cycle inversion to generate other conformers. In particular, a given chair form (e.g., IVa) can generate a second chair form (e.g., IVb). For cyclohexane itself, the two forms are indistinguishable. However, a careful inspection reveals that the inversion process changes all equatorial bonds to axial bonds, and vice versa. In the next section we see that for substituted cyclohexanes the two chairs are distinct isomers.

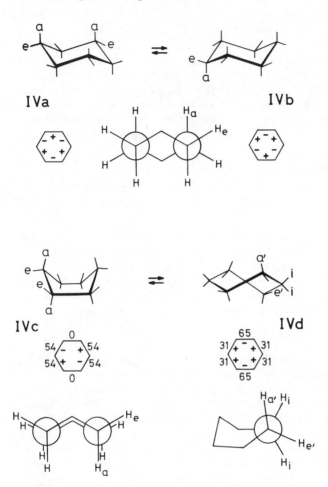

The reversible IVa/IVb cycle inversion is a complex process which can involve several transition states. The interested reader is referred to the outstanding review of Bucourt [5] for a detailed and clear description of the inversion process using the torsion angle representations. Let us simply state that the two chair forms are separated by a barrier of about 44 kJ/mol (10.5 kcal/mol).

Besides the rigid chair conformers, cyclohexane also exists as flexible forms which include the boat conformer IVc and the twist form IVd (or skew-boat). In fact, the flexible form gives rise to an infinite number of conformations by continuous variation of the torsion angles. The boat and twist forms represent the local energy maxima and minima, respectively, of the flexible form, with the former about 25 kJ/mol (6 kcal/mol), and

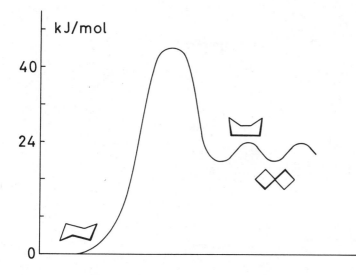

FIGURE 1   The conformational behavior of cyclohexane.

the latter about 21 kJ/mol (5 kcal/mol) above the global energy minimum of the chair forms.   Figure 1 shows the well-known plot of energy versus conformation for cyclohexane [1].

The boat conformation of cyclohexane, like the chair form, shows only axial and equatorial exocyclic positions [7].   For the twist conformations, however, pseudoequatorial (e'), pseudoaxial (a') and isoclinal (i) positions have been discriminated [9].

Increasing the number of methylene groups in alicyclic systems results in the possibility of additional conformers [1,5,10,11].   Without going into too many details, let us simply consider cycloheptane.   This molecule has two families of forms interconvertible by pseudorotation. These are the chair (Va) and the twist-chair (Vb) forms, as well as the boat (Vc) and the twist-boat (Vd) forms.   The twist-chair conformation is the most stable one; it has three different equatorial and axial positions, and two identical isoclinal (i) positions [1,5,12,13].

Introduction of a double bond into an alicycle considerably flattens the molecule.   Taking cyclohexene as an example, carbon atoms 1, 2, 3, and 6 have long been considered to be exclusively coplanar.   In this situation, the most stable forms are the two enantiomeric half-chair conformers VIa and VIe.   These interconvert via a transition state, which is the boat form VIc (also called half-boat), having an energy content of about 25 kJ/mol

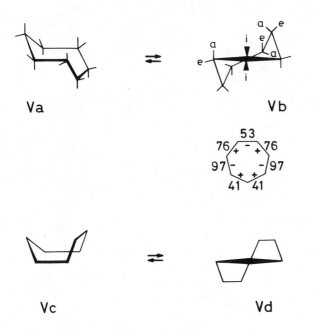

Va ⇌ Vb

Vc ⇌ Vd

(6 kcal/mol) above the half-chair energy. Another conceivable transition state is the fully planar conformer whose involvement, however, is unlikely due to a high energy of 39 kJ/mol (9.3 kcal/mol [1,5].

A slight twisting of the double bond has recently been calculated to be energetically possible. The resulting so-called sofa forms (VIb and VId) are just 3.3 kJ/mol (0.8 kcal/mol) above the global minimum. However, these forms do not correspond to a local energy minimum, but represent a passing conformation in the half-chair-to-boat interconversion path (VIa to VIe) [5].

10.2 Monosubstituted Carbocycles

As pointed out in the previous section, two isomeric monosubstituted cyclohexane derivatives are possible, namely, VIIa (with the substituent X being equatorial) and VIIb (X axial). The barrier between the two forms is usually in the range of 42 to 50 kJ/mol (10 to 12 kcal/mol), and the energy difference between the chair and boat forms does not differ markedly from that in unsubstituted cyclohexane. Many monosubstituted cyclohexane derivatives, like cyclohexane itself, preferentially adopt a chair conformation, and, more precisely, the conformation with the substituent in an equatorial position (VIIa).

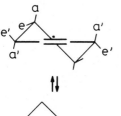

 VIa

 VIb

 VIc

 VId

 VIe

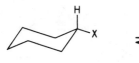

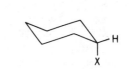

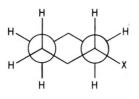

VIIa

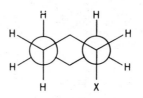

VIIb

Classically, two steric arguments are put forward to explain the equatorial preference. First, an axial substituent shows larger diaxial 1/3 interactions with the two hydrogen atoms (VIIIb) than does a 1-axial hydrogen (VIIIa); this is due to the larger steric bulk of the substituent as compared to a hydrogen atom. Also, an axial substituent experiences gauche interactions with C-3 and C-5 (VIIb, VIIIb). By occupying an equatorial position, the substituent becomes trans to C-3 and C-5 (VIIa, VIIIa) [1].

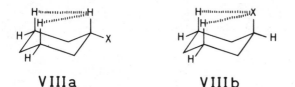

VIIIa   VIIIb

In accordance with the van der Waals radius of hydrogen (Section 3.3), the proposal has been made that hydrogen/hydrogen gauche interactions play a marked role in the conformational behavior of cyclic systems [14]. This proposal is based on the hypothesis that for a hydrogen atom, the equatorial position is more sterically hindered than the axial position. The axial preference of the hydrogen atom would leave a substituent with the equatorial position by default. The current evidence indicates that the "equatorial hydrogen effect" contributes only a minor amount to the conformational energy difference [15] (see also Section 9.2).

The energy difference between conformers VIIa and VIIb (equatorial and axial substituents, respectively) varies with the nature of the substituent. A wealth of relevant data has been obtained by theoreticians and experimentalists; a table of conformational energies published in 1967 reports many results found in the literature [16]. Selected values are presented in Table 2. The conformational energies of many other substituents have been assessed since 1967 and can be found in recent publications.

Less information is available on other monocyclic systems. It has been indicated, however, that in monosubstituted cyclobutanes and cyclopentanes the preferred position of a substituent is equatorial [1,17]. In cycloheptane, substituents at the isoclinal and equatorial positions have practically the same conformational energy [13].

Monosubstituted cycles including a sp^2 carbon call for a special discussion. An example of such compounds is cyclohexanone (IX), where minor ring distorsions exist as compared to cyclohexane. Like the latter molecule, and as opposed to monosubstituted cyclohexane derivatives, the two chair conformations (IXa and IXb) are identical. Because the barrier of rotation about a sp^2-sp^3 carbon-carbon bond is intrinsically lower than about a sp^3-sp^3 bond (see Section 9.4) cyclohexanone shows increased

TABLE 2 Standard Free Energy Change for the Axial/Equatorial
Equilibrium in Monosubstituted Cyclohexane Derivatives[a]

	$-\Delta G°$	
Substituent	kJ/mol	kcal/mol
F	0.63	0.15
Cl	1.80	0.43
Br	1.59	0.38
I	1.80	0.43
CN	0.71	0.17
CH_3	7.12	1.70
CH_2CH_3	7.33	1.75
$CH(CH_3)_2$	9.00	2.15
Phenyl	12.6	3.0
COOH	5.65	1.35
COO^-	8.04	1.92
$COOCH_3$	5.32	1.27
OH (Aprotic solvents)	2.18	0.52
(H-bond donor solvents)	3.65	0.87
OCH_3	2.51	0.60
$OCOCH_3$	2.51	0.60
NH_2 (Aprotic solvents)	5.03	1.20
(H-bond donor solvents)	6.70	1.60
NH_3^+	7.95	1.90
NO_2	4.61	1.10

[a]From Hirsch [16].

flexibility in the part of the ring containing the carbonyl group [5]. The
barrier of transition of the chair inversion has been calculated to be approximately 25 kJ/mol (6 kcal/mol) (experimental value around 21 kJ/mol) [5].

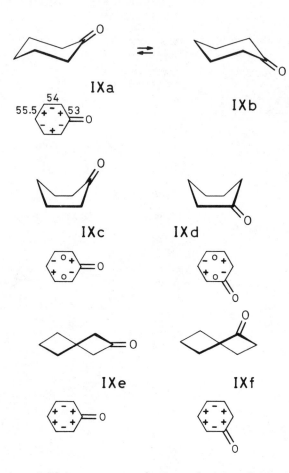

Cyclohexanone can further adopt two distinct boat conformations, one having a plane of symmetry (C_S boat, IXc) and the other devoid of it (C_1 boat, IXd). Two distinct twist conformations can also exist (IXe and IXf), both devoid of a plane of symmetry. These conformers IXc, IXd, IXe, and IXf have been calculated to have energies of approximately 22.6, 16.8, 13.4, and 16.8 kJ/mol (5.4, 4, 3.2, and 4 kcal/mol) above that of the chair form [5].

10.3 Bi- and Polysubstituted Carbocycles

Bi- and polysubstituted cyclic systems differ from monosubstituted cycles in that configurational aspects are involved. Consider, for instance, 2-methylcyclopropanol (X); this molecule contains two asymmetric carbons

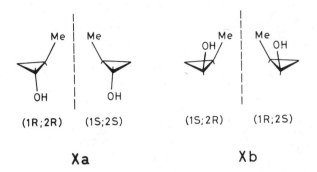

(1R;2R)   (1S;2S)   (1S;2R)   (1R;2S)

Xa   **Xb**

and exists as four stereoisomers in accordance with the rules discussed for acyclic systems (Section 6.1). The two enantiomeric trans-isomers (Xa) have the two substituents on either side of the plane of the cycle, whereas the two enantiomeric cis-isomers (Xb) have the substituents on the same side. Thus, each of the four stereoisomers of X is enantiomeric with one, and one only, of the three other isomers, and diastereoisomeric with the remaining two.

The same stereochemical characteristics exist for 1,2-disubstituted carbocycles having two different substituents (XI, R ≠ R'). Generalizing further, we consider a disubstituted carbocycle where the two different substituents R and R' are carried by any two carbon atoms of the cycle (XII). In such a situation, the condition for the existence of two centers of chirality and four stereoisomers is that the two branches of the cycle differ, i.e., n ≠ m. In the case of n = m, the molecule contains a plane of symmetry which causes the two enantiomeric pairs to degenerate; such molecules have only two achiral stereoisomers, namely a cis- and a trans-form, which are σ diastereoisomers. In the case of geminal substituents (1,1-disubstituted carbocycles) (XIII), a plane of symmetry also exists.

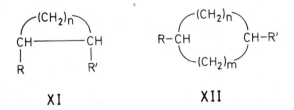

XI   **XII**

Let us now consider disubstituted carbocycles in both their configurational and conformational aspects by taking cyclohexane derivatives as examples. Achiral 1,1-disubstituted cyclohexanes (XIII) exist as two chair

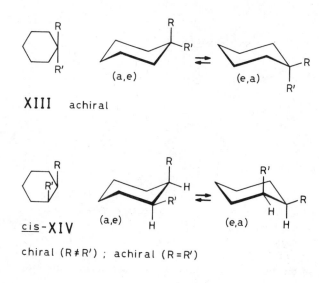

XIII achiral

cis-XIV (a,e) (e,a)

chiral (R≠R') ; achiral (R=R')

trans-XIV chiral

conformers. 1,2-Disubstituted derivatives (XIV) may exist as four stereoisomers each having two distinct chair conformations. The same is true
for 1,3-disubstituted derivatives (XV). Note that only one of the two
enantiomers of cis-XIV, trans-XIV, cis-XV, and trans-XV is represented.

The planar projection formulas of cyclohexane derivatives as shown
in structures XIII-XVI can obviously not be considered as correct approximations of the actual three-dimensional geometries of the molecules.
Such planar representations are useful, however, to symbolize and to
count the various possible stereoisomers [18].

In structures XIII-XVI, only chair conformations have been drawn,
although it is known that a few cyclohexane derivatives preferentially adopt
nonchair conformations, e.g., cyclohexane-1,4-dione, trans-1,3- and cis-
1,4-di-tert-butylcyclohexane [9]. With regard to the chair conformations
XIII-XVI, one must discriminate between compounds which experience a
diequatorial/diaxial equilibrium (trans-1,2-, cis-1,3-, and trans-1,4-
disubstituted cyclohexanes) and those which experience an axial-equatorial/

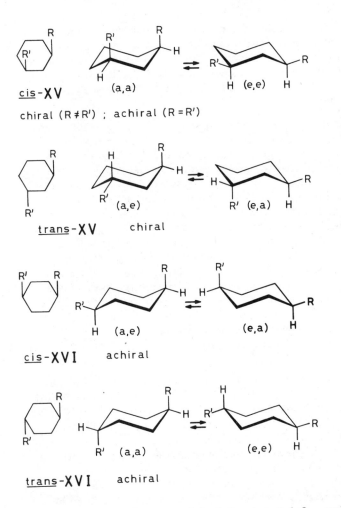

cis-**XV** (a,a) (e,e)

chiral (R≠R') ; achiral (R=R')

trans-**XV** chiral (a,e) (e,a)

cis-**XVI** achiral (a,e) (e,a)

trans-**XVI** achiral (a,a) (e,e)

equatorial-axial equilibrium (cis-1,2-, trans-1,3-, and cis-1,4-disubstituted cyclohexanes). In the latter derivatives, the balance influence of both substituents will affect the conformational behavior.

In the case of a diequatorial/diaxial equilibrium, each substituent considered separately tends to adopt an equatorial position (see Table 2, Section 10.2). In these molecules, however, an additional effect arises from the fact that the interactions between the substituents themselves are different in the diaxial and in the diequatorial conformations. As a result, substituent interactions also contribute to the overall conformational energy. At the present time, much remains to be studied regarding these interactions [19,20]. In trans-1,2-dichlorocyclohexane for example, the Cl/Cl gauche interaction destabilizes the diequatorial conformation [20].

The substituent interactions contain a steric and an electronic term [20] whose relative contributions are a function of the substituents involved, their relative positions, and the solvent. Thus, trans-1,2-, cis-1,3-, and trans-1,4-dichlorocyclohexane in CS_2 have diaxial/diequatorial energy differences of 1.97, 6.3 (or larger), and 0.17 kJ/mol, respectively (0.47, 1.5, and 0.04 kcal/mol) [21].

Disubstituted cyclohexanes having two identical R and R' substituents differ in some of their stereochemical characteristics from their analogs with different substituents. This is true for the cis-1,2- and cis-1,3-derivatives which now possess a plane of symmetry and are achiral (meso-isomers). The two chair conformers of the 1,1-, cis-1,2-, trans-1,3-, and cis-1,4-disubstituted cyclohexanes (R = R') always exist in a 50:50 ratio since they have exactly the same energy; in the case of 1,1-, trans-1,3-, and cis-1,4-derivatives, the two interconverting chairs are indistinguishable (degenerate interconversion). In the case of the achiral cis-1,2-derivative, the (a,e)- and (e,a)-chairs share an enantiomeric relationship. This compound affords an interesting example of a conformational (±)-pair. The identical energy levels of the two enantiomeric conformers forbid the apparition of ex nihilo chirality in this achiral molecule. The characteristics of the cis-1,2-derivative differ from those of the trans-1,3-derivative, which has configurational chirality and, therefore, cannot have conformational chirality (since any chiral structure has one, and only one, enantiomer).

When considering polysubstituted carbocycles (tri-, tetra-, penta-, hexasubstituted, etc.), the configurational and conformational possibilities become quite large. The designation of absolute stereochemistry is done according to the sequence rule procedure (Section 5.4). The relative stereochemistry can be specified by the modification of the sequence rule described in Section 6.3. However, the most satisfactory description of the relative configuration of polysubstituted monocycles is that recommended by the IUPAC [7]. According to these rules, the lowest numbered substituent is chosen as reference and is designated by r, followed by a hyphen, before its locant. The relation of the other substituents relative to the reference substituent is then expressed by adding c or t (for cis or trans) before their locant. For example, compound XVII and its enantiomer are c-2-amino-t-4-hydroxy-r-1-cyclohexanecarboxylic acid; similarly, XVIII and XIX are c-2,c-5-dimethyl-r-1-cyclopentanol and t-2,t-5-dimethyl-r-cyclopentanol, respectively. Compound XX is r-1,t-2,c-4-trichlorocyclopentane; the alternative direction of numbering would yield r-1,t-2,t-4, and, in such cases the IUPAC recommends choosing that numbering which gives a cis-attachment at the first point of difference. Additional rules are presented for special cases [7].

The chirality or achirality of polysubstituted cycles is not always easy to ascertain. Using the approach discussed in Chapter 6, it may be useful

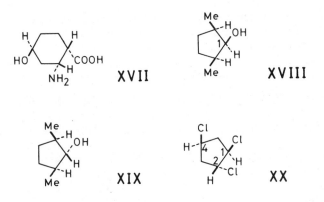

to search the molecule under consideration for constitutional symmetry; this is the case when a plane of symmetry bisects an endocyclic atom or bond (see Section 6.2). Constitutionally symmetrical molecules are achiral _meso_-forms with or without a pseudoasymmetric atom. For example, XVII and XX are chiral, while XVIII and XIX are 1-s-_meso_ and 1-r-_meso_ diastereoisomers, respectively.

The conformational aspects of polysubstituted cycles are complex because so many parameters are involved; these require a fair understanding of all attractive and repulsive interactions within the cycle itself, between substituents, and between substituents and the cycle. A detailed discussion of the conformational behavior of polysubstituted cyclohexanes is given by Hanack[1].

10.4 Nonsubstituted Heterocycles

Replacing a carbon atom with a nitrogen or oxygen atom in a saturated cycle has minute influence on the geometry of the cycle. Indeed, C-C bond distances show little difference with C-N or C-O distances (Table 1, Section 3.1); also, C-C-C, C-N-C, and C-O-C bond angles are usually in the range of $112° \pm 0.5°$.

Replacing a carbon with a sulfur, phosphorus, or silicon atom has a more profound influence due to the relatively long bond distances (Table 1, Section 3.1), and, in the case of sulfur to the relatively closed C-S-C angle (about $100°$; see also Section 3.2). Non-negligible cyclic distortions are thus expected when comparing heterocycles containing these atoms to their analogous carbocycle.

Baeyer strain in O- and N-containing heterocycles will be essentially the same as in the corresponding carbocycle, while it may be larger in sulfur-containing heterocycles. Bond opposition strain of gauche and eclipsed conformations (Pitzer strain) will differ between carbo- and heterocycles, as predicted from the rotation barriers of n-propane and heteroanalogs (Table 5, Section 9.3). Also, solvation energies in some cases may be greater for hetero- than for carbocycles, and, as a result, the conformational behavior of the former may show marked solvent dependency.

It appears that six-membered rings afford the most useful comparison between hetero- and carbocycles, and most of the following discussion will be devoted to them. When the endocyclic atom X in system XXI is either oxygen (oxane) or sulfur (thiane), the process of chair-chair ring reversal converts the unsubstituted ring into its superimposable mirror image, while all axial and equatorial exocyclic positions are interchanged. The two chair conformations are indistinguishable, i.e., they have exactly the same energy. The same holds true for silinane (X = SiH$_2$), which is identical to cyclohexane with respect to the number of exocyclic positions.

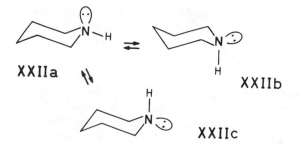

XXI

The behavior of piperidine (XXI; X = NH) differs from that of the above compounds since chair-chair ring reversal results in two diastereoisomeric conformers having the 1-H in an equatorial or axial position (XXIIa and XXIIb, respectively). The two diastereoisomeric chair conformations can also interconvert by nitrogen inversion (e.g., XXIIa - XXIIc), since the two representations XXIIb and XXIIc are identical. Ring reversal and nitrogen inversion are competitive processes and are often difficult to distinguish experimentally. The fact that both processes generate diastereoisomeric conformers is a further argument not to consider low-energy nitrogen inversion as a configurational, but as a conformational process (see Section 4.1).

The question of the preferred conformation of piperidine has received much attention and, as a consequence, has generated conflicting results

XXIIa

XXIIb

XXIIc

TABLE 3 Free Energy of Activation for Ring Inversion in Six-Membered Cycles[a]

Cyclic compound	$\Delta G^{\ddagger}$		
	kJ/mol	kcal/mol	(T°C)
Cyclohexane (X = CH$_2$)	43.2	10.3	(−67.0)
Piperidine (X = NH)	43.6	10.4	(−62.5)
Oxane (X = O)	39.8	9.5	(−80.0)
Thiane (X = S)	37.7	9.0	(−93.0)
Selenane (X = Se)	34.4	8.2	(−105.0)

[a]From Lambert and Featherman [23].

and quite a lot of confusion. The subject has been authoritatively summarized in quasiepic terms by Katritzky and collaborators [22], who conclude that the N-H equatorial form (XXIIa) is favored over XXIIb by 1.7 ± 0.8 kJ/mol (0.4 ± 0.2 kcal/mol) in the gas phase and in nonpolar solvents. An uncertainty remains with regard to polar solvents, in which the N-H axial form may well be favored [22].

The transition state for chair-chair ring inversion is generally recognized as being a half-chair conformation, three of which are conceivable for a pentamethylene heterocycle (XXIIIa-c) [23]. When the C-X torsional barrier is lower than the C-C one (Table 5, Section 9.3), the transition state XXIIIa is expected to be preferred because the heteroatom relieves the greatest amount of Pitzer strain. But, if the C-X torsional barrier is higher, then the transition state XXIIIc is preferred since it places the heteroatom in the least eclipsed portion of the ring [23]. Globally, the transition barriers appear to depend directly on the magnitude of the C-X torsional energy, and the adopted transition state is just one of many factors controlling this transition barrier. If we look at the free energy of activation of the ring inversion process as determined by NMR analysis (Table 3) and compare it with the C-X torsional barriers as listed in Section 9.3, Table 5, a direct correlation is evident [23].

XXIIIa　　　　**XXIIIb**　　　　**XXIIIc**

In the case of unsubstituted six-membered cycles containing two heteroatoms, numerous compounds have been studied, some of which are discussed by Hanack[1]. Let us briefly review piperidine analogs (XXIV) where the second heteroatom is O or N[22]. Introduction to this second heteroatom in the 4-position to yield morpholine (XXIVa, Z = O) or piperazine (XXIVa, Z = NH) has a limited influence on the conformational equilibrium. The two heteroatoms are remote from each other, and their mutual interactions will remain minor. In particular, the N-H equatorial form appears to predominate[22].

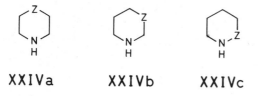

XXIVa   XXIVb   XXIVc

When the second heteroatom is in the 3-position, considerable interaction exists between the two geminal, electron-rich centers, which has a profound conformational influence. In tetrahydro-1,3-oxazine (XXIVb, Z = O) and hexahydropyrimidine (XXIVb, Z = NH), strong N-H axial preference has been found (XXVb). This preference was attributed to a diaxial repulsion between two electron lone pairs in conformation XXVa, a crude approximation considering that the lone pairs are more diffuse than suggested in XXV. More recently, it has been proposed that the N-H axial preference arises from an attractive gauche interaction between the polar N-H axial bond and the C(2)-Z bond[22]. The problem of dipole-dipole interactions for geminal electron-rich centers will be mentioned again when discussing the anomeric effect in substituted heterocycles (Section 10.5).

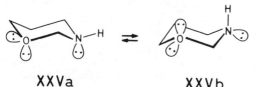

XXVa   XXVb

In tetrahydro-1,2-oxazine (XXIVc, Z = O) the N-H equatorial conformer is favored, apparently due to minimization of lone pair gauche interactions, or to maximal NH-O lone pair overlap[22].

Unlike six-membered heterocycles, but like cyclopentane itself, five-membered heterocycles are pseudorotational systems assuming a continuous set of conformations. The heteroatom(s) can occupy distinct positions in the half-chair and envelope conformations, but the absence of well-

TABLE 4 Barrier of Inversion in Four-Membered Cycles[a]

Cyclic compound XXVI	Barrier	
	kJ/mol	kcal/mol
Cyclobutane (X = CH_2)	6.2	1.48
Siletane (X = SiH_2)	5.28	1.26
Azetidine (X = NH)	5.28	1.26
Oxetane (X = O)	0.4	0.1
Thietane (X = S)	3.14	0.75

[a]From Moriarty [17].

XXVI

defined maxima and minima of energy leaves only theoretical interest to the discrimination of conformers. (A relevant collection of structural data has been published some years ago) [24].

Four-membered heterocycles have also been the center of much interest. It was found that the barrier of ring reversal when going from one puckered form to the other is quite sensitive to the nature of the heteroatom. While siletane (XXVI, X = SiH_2) and azetidine (XXVI, X = NH) have barriers comparable to that of cyclobutane, the barrier is low in thietane (XXVI, X = S) and especially in oxetane (XXVI, X = O) (Table 4) [17]. The barrier is thus significantly lower when one ring position bears no exocyclic atoms. The smaller C-S-C valency angle as compared to C-O-C destabilizes the planar form and increases the barrier. Further, the two puckered forms of azetidine are nonequivalent (axial or equatorial N-H), the energy difference being 1.13 kJ/mol (0.27 kcal/mol) [17].

10.5 Substituted Heterocycles

In substituted heterocycles, the interaction between the endocyclic heteroatom(s) and the substituent(s) may have marked conformational consequences. Selected cases will be considered in this section.

Nitrogen-containing heterocycles may carry N-substituents whose stereochemical characteristics differ from those of substituents on other ring positions. The general configurational lability of nitrogen atoms allows the generation of diastereoisomeric conformations without ring reversal. However, this is not a general rule, as several cases of configurational stability of nitrogen are known. Table 5 lists selected examples of nitrogen inversion in three- to seven-membered heterocycles. On the surface, it appears that the barrier of nitrogen inversion is controlled by ring size; indeed, three-membered rings have high-energy barriers which result in genuine configurational stability for oxaziridines and diaziridines. But many factors influence the barrier, e.g., the steric bulk and electronegativity of the N-substituent, the electronic conjugations within the cycle as modified by the presence of a second heteroatom, and others [25]. For saturated six-membered heterocycles, the influences of the molecular environment have been rationalized in terms of increments on N-methyl inversion barrier [26].

When comparing barriers of nitrogen inversion and of ring reversal, it is interesting to note that in four- and five-membered rings the latter process requires less energy than the former. In azetidines, for example, the two conformers will interconvert preferentially by ring reversal. But, in six-membered rings the two processes may be competitive due to the comparable energy levels.

The preferred position of a N-substituent is usually found to be the equatorial one. In N-methylpiperidine, for example, recent investigations have shown the N-methyl group to prefer the equatorial orientation by 10.5 to 12.6 kJ/mol (2.5 to 3.0 kcal/mol) [27], the latter value being that measured in an apolar solvent [28]. These values are considerably higher than the energy difference reported for methylcyclohexane (see Table 2). In N, N'-dimethylpiperazine, the free energy difference between an axial and an equatorial N-methyl group has been found to be 12.1 to 12.6 kJ/mol (2.9 to 3.0 kcal/mol), a result confirming the above values [27]. In C-methylated N-methylpiperidines, the N-methyl group somewhat influences C-methyl groups, in particular, by decreasing the conformational energy of a 2-methyl group [29]. The latter example affords another illustration of conformational interactions in ring substituents, as already mentioned in Section 10.3.

In the case of oxygen- or sulfur-containing heterocycles, some electron-rich substituents adjacent to carbon-2 show an unusual conformational behavior. Indeed, an halogen atom or an oxygen- or sulfur-containing substituent will tend to markedly prefer an axial position when vicinal to an endocyclic oxygen (or sulfur) atom. This axial preference is known as the anomeric effect, suggesting its significance in the conformational control of carbohydrates [30].

TABLE 5 Free Energy of Activation of Heterocyclic Nitrogen Inversion[a]

Compound	$G^{\ddagger b}$		
	kJ/mol	kcal/mol	(T°C)
Aziridine derivatives			
N–R **XXVII**			
R = CH_2CH_3	81.3	19.4	(108)
R = $C(CH_3)_3$	71.2	17.0	(52)
R = Phenyl	49.0	11.7	(−40)
R = $CON(CH_3)_2$	41.5	9.9	(−86)
R = $COOCH_3$	29.7	7.1	(−138)
R = $SO_2C_6H_5$	52.0	12.4	(−30)
Oxaziridine derivatives			
O N–R **XXVIII**			
R = $C(CH_3)_3$	138.0	33.0	(120)
Diaziridine derivatives			
R N N–R' R'' R''' **XXIX**			
R = R'' = CH_3; R' = R''' = Benzyl	114.0	27.3	(70)

TABLE 5   (continued)

Compound	$G^{\pm b}$		
	kJ/mol	kcal/mol	(T°C)

Azetidine derivatives

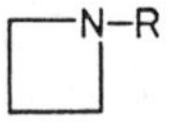

XXX

R = CH$_3$	4.27	10.2	(-69)
R = Cl	~56.0	~13.4	(-20)
N-Amino-3,3-dimethylazetidine	42.7	10.2	(-65)

Pyrrolidine derivatives

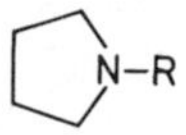

XXXI

R = CH$_3$	35.2	8.4	(-98)
R = Cl	43.2	10.3	(-68)
N-Hydroxy-3,3-dimethylpyrrolidine	54.5	13.0	(-11) (CDCl$_3$)
	63.8	15.0	(30) (D$_2$O)

1,2-Oxazolidine derivatives

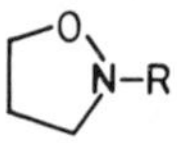

XXXII

R = CH$_3$	65.4	15.6	(42) (CDCl$_3$)
	70.8	16.9	(62) (D$_2$O)
R = CH(CH$_3$)$_2$	62.0	14.8	(5)
R = CH$_2$OCH$_3$	43.2	10.3	(-74)

TABLE 5 (continued)

Compound	$G^{\ddagger b}$		
	kJ/mol	kcal/mol	(T°C)
Six-membered rings			
cis-2,6,N-Trimethylpiperidine	61.2	14.6	(33) (H_2O)
N,N'-Dimethylpiperazine	45.7	10.9	(44) (H_2O)
N-Methyltetrahydro-1,2-oxazine	57.4	13.7	(5)
Hexahydroazepine derivatives			

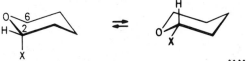

XXXIII

	kJ/mol	kcal/mol	(T°C)
R = CH_3	28.5	6.8	(-125)
R = Cl	38.5	9.2	(-87)

[a]From Lambert [25].

[b]No solvent or apolar solvent unless otherwise specified.

The magnitude of the anomeric effect has been defined as the free energy difference between the axial and equatorial conformer (XXXIVa and XXXIVb, respectively), plus the ordinary conformational preference of the 2-substituent (Section 10.2, Table 2) [24]. Using appropriately 2-substituted 4-methyltetrahydropyrans, the anomeric effect was found to be equal to 11.3 kJ/mol (2.7 kcal/mol) for chlorine, and larger than 12 kJ/mol for bromine and iodine (neat liquid). For hydroxy, alkoxy, and acyloxy groups, these values are smaller (3.8-5.9 kJ/mol, 0.9-1.4 kcal/mol) and more solvent-dependent than for the halogens [24].

XXXIVa   ⇌   **XXXIVb**

The reasons for the anomeric effect are electronic in nature and have led to many speculations. A simple, but incomplete approximation considers dipole-dipole repulsions as favoring the axial conformer. A somewhat incorrect, but suggestive representation of oxygen nonbonding electrons shows one gauche halogen-lone pair interaction in the axial conformer, but two such interactions in the equatorial form (XXXV) [24]. By realistically considering the oxygen lone pairs as nonequivalent, interaction diagrams were obtained which clearly indicate an axial X to be preferred [31]. Experimentally, it has consistently been found that the O-C(2) distance is shorter than the O-C(6) distance, indicating mixing of the oxygen nonbonding orbitals with the C(2)-H and C(2)-X antibonding orbitals. For symmetry reasons, this electron donation can be effective only from the axial lone pair into the C-X orbitals of an axial (anti) substituent X.

XXXV

The anomeric effect has also been studied using dioxanes, e.g., 2-substituted 1,3-dioxanes (XXXVI). Furthermore, the 1,3-dioxane ring system has proven to be fascinating in that it allows stereochemical investigations of nonbonded interactions between the 5-substituent and the two endocyclic oxygen atoms. Acid catalysis of 5-substituted 1,3-dioxanes leads to equilibrium between axial- and equatorial-X isomers [32]. Convenient molecules have proven to be the 2-isopropyl derivatives (XXXVII).

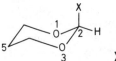

XXXVI

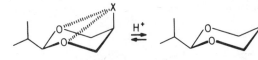

XXXVII

The position of equilibrium between the epimers XXXVII is mainly dependent upon the nature of the 5-substituent. As a general trend, electron-rich substituents such as Cl, Br, OCH_3, and SCH_3, show a marked preference for the equatorial position where repulsive interactions are minimized. In contrast, groups such as F, CH_2OH, NO_2, SO_2CH_3, and the positively charged substituents are predominantly axial [33]. While solvent effects are far from negligible, a comparison of the axial-equatorial free energy differences displayed by a given substituent attached either to 1,3-dioxanes or cyclohexane appears as a valuable means to investigate some nonbonded interactions [33].

Let us end this section by just mentioning the great promise shown by cyclic peptides as conformational models [34]. The simplest molecule in this series is diketopiperazine (XXXVIII), the cyclic dimer of glycine, but larger cycles are of easy access. The rings and their substituents can almost be tailored to wish, allowing the investigation of interactions between a large variety of functional groups [35].

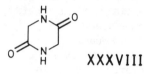

XXXVIII

10.6 Fused Ring Systems

Two fused rings have two endocyclic atoms in common, as opposed to bridged rings which have more than two common atoms (see next section).

Configurational and conformational aspects are apparent in the stereochemistry of fused bi- and polycyclic systems. Configurational isomerism results from the fusion of two alicyclic rings. Considering first the general case of two cycles of undefined size, it can be seen that the angular hydrogens can either be trans or cis. The two stereoisomeric molecules are termed trans and cis (XXXIX and XL, respectively), and may be represented in different ways as shown.

XXXIX

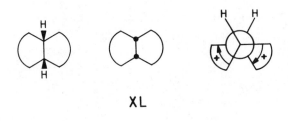

XL

The two carbon atoms engaged in ring fusion (i. e., at valley positions) may be achiral or asymmetric depending on the symmetry properties of the molecule. But, in either case, the cis-trans diastereoisomerism displayed by bicyclic fused systems is comparable to cases of diastereoisomerism in bisubstituted carbocycles (Section 10.3) and does not involve new stereochemical principles.

The two diastereoisomeric forms of decalin (trans-decalin, XLI, and cis-decalin, XLII) differ in their conformational behavior. If we consider only the lower energy chair-chair conformations, and not the higher energy boat-chair and boat-boat forms, it is proved that trans-decalin is conformationally rigid (XLI). But, cis-decalin exists as two enantiomeric conformers [(±)-pair of conformers] separated by a barrier of inversion ($\Delta G^{\ddagger}$ = 54 kJ/mol [12.9 kcal/mol] at -60°C) which is larger than that of cis-1,2-dimethyl-cyclohexane ($\Delta G^{\ddagger}$ = 41.9 kJ/mol [10.0 kcal/mol] at -60°C) and that of cyclohexane (Table 2) [36].

 XLI

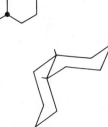 XLII

If, in decalin, either ring is considered as substituent of the other, the trans-isomer displays a diequatorial pattern, whereas the cis-isomer is axial-equatorial. Also, the latter isomer differs from the former by three gauche n-butane interactions. cis-Decalin, therefore, has a higher energy content than trans-decalin. Experimental values are in the range 8.8 to 11.3 kJ/mol (2.1 to 2.7 kcal/mol) [1]. It is important, however, to remember that the cis- and trans-diastereoisomers are configurational isomers, i.e., separated by a high-energy barrier involving C-C bond cleavage.

Hydrindane is an interesting molecule because it affords a system resulting from the fusion of two rings of different size, and also, because it represents a portion (rings C and D) of the steroid nucleus. The dissimilarity of the two rings causes the two carbons atoms engaged in ring fusion to be asymmetric. Since they carry identical ligands, the molecule will exist as two stable trans-enantiomers (XLIII) and a meso-cis-isomer (XLIV). As with the decalins, the trans-hydrindanes are conformationally rigid molecules, whereas cis-hydrindane is flexible and exists as a (±)-pair of conformers. The free energy of activation of the conformational inversion is about 26.8 kJ/mol (6.4 kcal/mol) at -130°C [37]. This value is considerably smaller than the inversion barrier of cis-decalin. Further, the cis- and trans-diastereoisomers of hydrindane have an energy difference of about 2.9 kJ/mol (0.7 kcal/mol) which is significantly smaller

XLIII

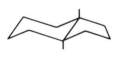

(R;R)   (S;S)

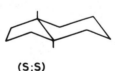

XLIV

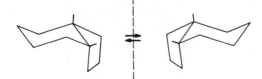

than that of the decalin diastereoisomers. These differences are due to
the influence of the five-membered ring and its more planar structure (as
compared to a six-membered ring) on the geometry and strain conditions
of the fused system. As a result, the barrier of conformational inversion
(for the cis-isomer) and the strain differences between diastereoisomers
are minimized in hydrindane as compared to decalin.

It must be realized that the cis-trans isomerism resulting from ring
fusion is impossible on steric grounds for the smallest rings (cyclopropane
and cyclobutane). Thus, only the cis-form of bicyclo [1.1.0] butane (XLV)
and bicyclo [2.2.0] hexane (XLVI) is known, the trans-juncture in these systems being rendered prohibitive by steric strain [17].

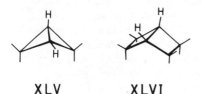

XLV   **XLVI**

The concept of torsion angle can be gainfully applied to the stereochemistry of ring junction [5]. A junction can thus be defined by two
torsion angles of junction (Φ and Φ', XLVII), i.e., the torsion angle of
each ring which has the common bond as its central bond [5]. When the two
torsion angles Φ and Φ' are of opposite sign, the junction is trans (e.g.,
XXXIX), whereas it is cis for angles of the same sign (e.g., XL).

XLVII

The merit of the torsion angle concept is to allow stereochemical differentiation of ring junction involving one trigonal carbon atom. Let us
consider $\Delta^{1(8a)}$-octalin (one of the structural isomers of octahydronaph-
thalene) (XLVIII). This molecule has one asymmetric carbon (the tetrahedral carbon engaged in ring fusion) and is chiral; this configurational
aspect, however, will not be discussed further. The conformational
flexibility of the molecule results in two distinct groups of conformers

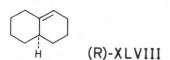

(R)-XLVIII

XLIXa   XLIXb

(XLIXa and XLIXb). Those conformers having torsion angles of junction of opposite signs (e.g., XLIXa) are termed quasi-trans by analogy with the above discussed examples. When the torsion angles are of the same sign (e.g., XLIXb), the conformers are designated quasi-cis. A priori, the quasi-cis conformers are expected to be less stable than the quasi-trans forms due to the marked opening of the torsion angle Φ' [5]. The concept of quasi-cis and quasi-trans conformational isomerism is of particular significance in the stereochemistry of steroids.

Tricyclic fused systems must be considered from two viewpoints, namely, the relationship between two adjacent (fused) rings, and the steric relationship between the two nonadjacent (external) rings. The perhydro-phenanthrenes (L) are interesting model compounds of great stereochemical richness. Four asymmetric carbon atoms are apparent (4a, 4b, 8a, 10a), and the molecule exists as 10 stereoisomers (two meso-forms and four (±)-pairs). The junction between rings A and B, and that between rings B and C, are described by the same cis/trans notation as discussed above. The relationship between rings A and C is described by considering the relative position of the hydrogens on C-4a and C-4b. The prefixes syn- and anti- are used when these hydrogens are on the same or opposite side, respectively. (IUPAC rules now recommend the terms cisoid and transoid to be used instead of syn and anti [7].) The two meso-forms are the cis-syn-cis and trans-syn-trans-isomer (La and Lb, respectively), since a plane of symmetry is evident. The four other forms (Lc-f) are chiral, and only one enantiomer of each (±)-pair is drawn [38].

The relative energy of perhydrophenanthrene stereoisomers is best discussed when considering the preferred conformations and their number of equatorial or axial bonds at the carbons engaged in ring fusion. The isomer of lowest energy is thus the trans-anti-trans-form (Le) which has four equatorial bonds. Isomers of higher energy (about 12 kJ/mol) are the

La

<u>cis</u>-<u>syn</u>-<u>cis</u>

Lb

trans-syn-trans

Lc

<u>cis</u>-<u>syn</u>-<u>trans</u>

Ld

cis-anti-trans

Le

trans-anti-trans

Lf

<u>cis</u>-<u>anti</u>-<u>cis</u>

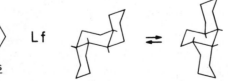

cis-syn-trans- (Lc) and cis-anti-trans-form (Ld) having three equatorial and one axial bond. Two isomers have two equatorial and two axial bonds; they are the cis-syn-cis- (La) and cis-anti-cis-form (Lf). The isomer of highest energy appears to be the trans-syn-trans-form (Lb), a rigid molecule in which the central ring is forced in a boat conformation [1].

The isomers of perhydrophenanthrene having none or one cis-junction are conformationally rigid; on the other hand, two cis-junctions render the molecule flexible. As a result of this conformational mobility, the cis-syn-cis-diastereoisomer exists as a racemic mixture of two enantiomeric conformers (La), while the cis-anti-cis-form isomerizes between two diastereoisomeric conformers (Lf).

The introduction of one or several substituents on a fused ring system renders their stereochemistry far more complex, both in configurational and conformational terms. Taking 2-substituted decalin (LI) as an example, one can see that the substituent desymmetrizes the molecule, making the two central atoms chiral. Three asymmetric carbon atoms thus exist in substituted decalins, resulting in eight stereoisomers. The diastereoisomers are designated as cis-cis, cis-trans, trans-cis, and trans-trans; the first term describes the type of ring fusion, while the second term describes the position of the substituent relative to the C(8)-C(9) bond. In other terms, R and C-8 are considered as substituents of ring A, and their relative position is designated as in the case of disubstituted monocycles (Section 10.3). One cis-cis-isomer is shown (LIa); like cis-decalin, it exists as a mixture of two conformers, which are no longer enantiomeric, but diastereoisomeric, due to the equatorial or axial position of the substituent. The former conformer can be expected to predominate.

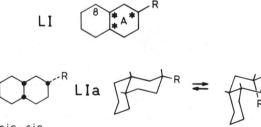

LI

LIa

cis-cis

For cyclic compounds which are drawn according to a well-established orientation, bonds projecting behind the plane are designated α (e.g., LII, R) while those projecting to the front are designated β (e.g., LII, R'). A substituent of unknown configuration is represented by a wavy line and designated ζ (xi) (e.g., LII, R'').

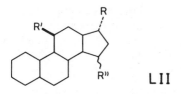

LII

For steroids (LII) and terpenes, the designations α and β represent absolute configurations. They build a specialized but useful nomenclature complementary to the general nomenclatures discussed in Chapter 5.

10.7 Bridged Ring Systems

Bridged ring systems are often synthetized as "rigid" analogs of simpler cyclic or acyclic molecules. By blocking various functional groups in selected relative positions, they allow a rational approach of topology-dependent chemical and biochemical properties. However, the conformational rigidity of bridged rings is a relative notion. Only the smaller, highly strained homologs are genuinely rigid, since in these molecules ring deformations originate only from vibrations in bond lengths and angles. A minor extent of conformational freedom becomes apparent in norbornane (bicyclo[2.2.1]heptane, LIII); this molecule has strict C_{2v} symmetry, but variously substituted derivatives exist in twisted conformations (LIIIb and LIIIc) [39]. Although the angles of twist measured (dihedral angle variations) are only minor (some degrees) [39], they are of interest because they imply that a slight distorsion of the C_{2v}-conformer requires little energy.

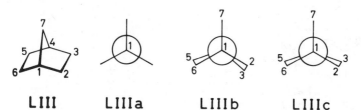

LIII LIIIa LIIIb LIIIc

Bicyclo[2.2.2]octane (LIV) in its totally eclipsed form (LIVa) has D_{3h} symmetry. Both theoretical calculations and experimental results indicate that the molecule possesses a broad energy minimum for twisting around the C(1)-C(4) axis [2,40,41]. The resulting twisted conformer LIVb and its enantiomer have D_3 symmetry; the estimations of the angle of twist vary between 7° and 15°, and the twisted conformer appears to be favored by only 0.4 kJ/mol (0.1 kcal/mol).

LIV   **LIVa**   **LIVb**

As compared to the above molecules, the conformational freedom is increased in bicyclo 3.2.1 octane (LV), bicyclo 3.3.1 nonane (LVI) and similar molecules. In the latter compound, the chair-chair conformer (LVIb) is favored over the boat-boat (LVIa) by 6.3 to 10.5 kJ/mol (1.5 to 2.5 kcal/mol) [2].

 LVa

 **LVb**

LVIa   **LVIb**   **LVIc**

Bridged ring systems display also interesting configurational characteristics. A favorite ring system is norbornane (LIII), which appears as a cyclohexane ring forced in a strained boat conformation by a methylene bridge. A substituent adjacent to carbons 2, 3, 5, or 6 is designated exo or endo, respectively, depending on its equatorial or axial position relative to the boat skeleton (LVII) (see also IVc, Section 10.1). The position of a substituent on carbons 1 and 4 is unequivocal, and is designated bridgehead.

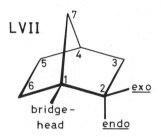

LVII

bridge-
head   endo   exo

TABLE 6 Stereoisomerism of Disubstituted Norbornane Derivatives
(Two Different Substituents)

Positions of substitution[a]	Number of stereoisomers[b]
1,4	1 Achiral configuration
7,7 (Geminal)	1 Achiral configuration
2,2 (Geminal)	2 Pairs of enantiomers
1,7	1 Pair of enantiomers
1,2 (Also 1,3)	2 Pairs of enantiomers
2,7	4 Pairs of enantiomers (exo-syn, endo-syn, exo-anti, endo-anti)
2,3 (Also 2,5 and 2,6)	4 Pairs of enantiomers (exo-exo, exo-endo, endo-exo, endo-endo)

[a] See Structure LVII.
[b] Permutation of the two substituents is not considered since it resorts to constitutional isomerism.

Considerable configurational restriction exists in norbornane and analogous systems. Norbornane monosubstituted in the 2-position has three asymmetric carbons, but for steric reasons carbons 1 and 4 behave as a single element of chirality. Four stereoisomers exist for such derivatives, namely, (+)- and (-)-endo, and (+)- and (-)-exo. Norbornane 1- or 7-monosubstituted does not display stereoisomerism. It is obvious that these various behaviors, in terms of symmetry, arise from the non-identity of positions 7 (bridge), 1 or 4 (bridgehead), and 2, 3, 5, or 6 (peripheral), respectively. These various positions result in the great stereochemical complexity of disubstituted norbornane derivatives. In order to reveal this complexity, Table 6 presents a correlation between the number of stereoisomers and the pattern of disubstitution. Examination of molecular models is suggested as the best if not only way of fully comprehending Table 6.

References

1. M. Hanack, Conformation Theory, Academic Press, New York, 1965.

2. N. L. Allinger, M. T. Tribble, M. A. Miller, and D. H. Wertz, J. Am. Chem. Soc., 93:1637 (1971).

3. F. A. Cotton and B. A. Frenz, Tetrahedron, 30:1587 (1974).

4. D. Cremer, J. Am. Chem. Soc., 99:1307 (1977).

5. R. Bucourt, The torsion angle concept in conformational analysis, in Topics in Stereochemistry, vol. 8 (edited by E. L. Eliel and N. L. Allinger), Wiley, New York, 1974, pp. 159-224.

6. D. Cremer and J. A. Pople, J. Am. Chem. Soc., 97:1358 (1975).

7. IUPAC Commission on Nomenclature of Organic Chemistry, 1974 recommendations for section E, fundamental stereochemistry, Pure Appl. Chem., 45:11 (1976).

8. O. Hassel, Tidssk. Kjemi Bergv. Metal., 3:32 (1943).

9. G. M. Kellie and F. G. Riddell, Non-chair conformations of six-membered rings, in Topics in Stereochemistry, vol. 8 (edited by E. L. Eliel and N. L. Allinger), Wiley, New York, 1974, pp. 225-269.

10. F. A. L. Anet, in Conformational Analysis (edited by G. Chiurdoglu), Academic Press, New York, 1971, pp. 15-29.

11. J. Sicher, The stereochemistry of many-membered rings, in Progress in Stereochemistry, vol. 3 (edited by P. B. D. de la Mare and W. Klyne), Butterworths, London, 1962, pp. 202-263.

12. K. von Bredow, H. Friebolin, and S. Kabuss, in Conformational Analysis (edited by G. Chiurdoglu), Academic Press, New York, 1971, pp. 51-58.

13. M. Christl and J. D. Roberts, J. Org. Chem., 37:3443 (1972).

14. D. H. Wertz and N. L. Allinger, Tetrahedron, 30:1579 (1974).

15. E. Ōsawa, J. B. Collins, and P. v. R. Schleyer, Tetrahedron, 33:2667 (1977).

16. J. A. Hirsch, Table of conformational energies - 1967, in Topics in Stereochemistry, vol. 1 (edited by N. L. Allinger and E. L. Eliel), Wiley, New York, 1967, pp. 199-222.

17. R. M. Moriarty, Stereochemistry of cyclobutane and heterocyclic analogs, in Topics in Stereochemistry, vol. 8 (edited by E. L. Eliel and N. L. Allinger), Wiley, New York, 1974, pp. 271-421.

18. J. E. Leonard, G. S. Hammond, and H. E. Simmons, J. Am. Chem. Soc., 97:5052 (1975).

19. D. Voisin and B. Gastambide, Bull. Soc. Chim. France II, 375 (1975).

20. N. S. Zefirov, L. G. Gurvich, A. S. Shashkov, M. Z. Krimer, and E. A. Vorob'eva, Tetrahedron, 32:1211 (1976).

21. O. A. Subbotin and N. M. Sergeyev, Anal. Chem., 48:545 (1976).

22. I. D. Blackburne, A. R. Katritzky, and Y. Takeuchi, Acc. Chem. Res., 8:300 (1975).

23. J. B. Lambert and S. I. Featherman, Chem. Rev., 75:611 (1975).

24. C. Romers, C. Altona, H. R. Buys, and E. Havinga, Geometry and conformational properties of some five- and six-membered heterocyclic compounds containing oxygen or sulfur, in Topics in Stereochemistry, vol. 4 (edited by E. L. Eliel and N. L. Allinger), Wiley, New York, 1969, pp. 39-97.

25. J. B. Lambert, Pyramidal atomic inversion, in Topics in Stereochemistry, vol. 6 (edited by N. L. Allinger and E. L. Eliel), Wiley, New York, 1971, pp. 19-105.

26. I. J. Ferguson, A. R. Katritzky, and D. M. Read, J. Chem. Soc. Chem. Comm., 225 (1975).

27. F. A. L. Anet and I. Yavari, Tetrahedron Lett., 2093 (1976).

28. P. J. Crowley, M. J. T. Robinson, and M. G. Ward, Tetrahedron, 33:915 (1977).

29. E. L. Eliel and D. Kandasamy, Tetrahedron Lett., 3765 (1976).

30. H. Paulsen, Die Stärke - Starch, 27:397 (1975).

31. O. Eisenstein, Nguyen Trong Anh, Y. Jean, A. Devaquet, J. Cantacuzène, and L. Salem, Tetrahedron, 30:1717 (1974).

32. M. J. T. Robinson, Tetrahedron, 30:1971 (1974).

33. M. K, Kaloustian, N. Dennis, S. Mager, S. A. Evans, F. Alcudia, and E. L. Eliel, J. Am. Chem. Soc., 98:956 (1976).

34. C. M. Deber, V. Madison, and E. R. Blount, Acc. Chem. Res., 9:106 (1976).

35. D. B. Davies and M. A. Khaled, J. Chem. Soc. P.T.II, 1238 (1976).

36. D. K. Dalling, D. M. Grant, and L. F. Johnson, J. Am. Chem. Soc., 93:3678 (1971).

37. W. B. Moniz and J. A. Dixon, J. Am. Chem. Soc., 83:1671 (1961).

38. K. Mislow, Introduction to Stereochemistry, Benjamin, New York, 1966, pp. 99-100.

39. C. Altona, in Conformational Analysis (edited by G. Chiurdoglu), Academic Press, New York, 1971, pp. 1-13.

40. E. M. Engler, L. Chang, and P. v. R. Schleyer, Tetrahedron Lett., 2525 (1972).

41. O. Ermer and J. D. Dunitz, Helv. Chim. Acta, 52:1861 (1969).

11

The Concept of Prostereoisomerism

The previous chapters have been devoted to stereoisomeric relationships between distinct molecular entities, while steric (morphic and topic) relationships between molecular fragments have been briefly presented in Section 4.2.

In this chapter, we will discuss relationships between groups or atoms of same constitution within intact molecules (topic relationships). Internal comparison of these groups by symmetry operations allows a simplified description of molecular structure. Also, the molecular environments of constitutionally identical groups are examined. If these environments are stereoisomeric, then the molecule is said to possess elements of prostereoisomerism. Mislow and Raban [1] have given a definitive classification of steric relationships of groups in intact molecules; the following discussion is based on this classification.

11.1 Criteria of Prostereoisomerism: Homotopic Groups and Faces

Let us consider the two hydrogen atoms of 1,1-dichloroethylene (I). Their molecular environments are identical in every aspect. These

I

environments consist of a two-dimensional space, where the rest of the molecule appears in the sequence H-C-Cl-C-Cl. The two hydrogen atoms are said to be stereochemically equivalent (homotopic) and are indistinguishable by any experimental test [2]. However, the <u>molecular environment</u> is sometimes a tedious criterion of equivalence which is gainfully complemented by simple <u>symmetry</u> considerations. The latter show that groups are homotopic if they can be interconverted by rotation about an axis of symmetry C_n ($\infty > n > 1$). In molecule I, rotation about the C_2 axis interchanges the two hydrogen atoms and results in a structure indistinguishable from the original. It is easy to see that molecule I contains two other equivalent groups, namely, the two chlorine atoms.

II

In toluene (II), a rotation of 120° about the C(4)-C(7) axis interchanges the three hydrogen atoms. But these atoms can be considered equivalent only if free rotation of the methyl group is assumed. Toluene is an interesting model in this aspect because it shows that the time factor is of importance to the concept of group equivalence. If, in toluene, the methyl rotation is fast relative to the time scale of the means of observation, then "free rotation" is observed, and the three hydrogens appear equivalent. If methyl rotation is slow in the observer's time scale, equivalence is lost. For example, let us assume toluene to be "frozen" in the depicted conformation (II); it is apparent that the hydrogen atom located in the plane of the phenyl ring experiences a different molecular environment than the other two hydrogens, and that the former atom is not equivalent to the latter two atoms.

The molecular environment within a given molecule can be defined relatively to <u>faces</u> of the molecule instead of groups. For example, the planar molecule formaldehyde (III) has two faces. Since there is no way to distinguish between these two faces, they are said to be equivalent.

Besides symmetry and molecular environment criteria, the <u>substitution criterion</u> is useful to assess equivalence of groups or atoms. In this procedure, the groups under consideration are replaced in turn by a test group, e.g., H is replaced by D. The resulting structures are examined for identity. If they are indistinguishable (superimposable by rotation

III

and translation), the groups are equivalent. Thus, substituting either of the two hydrogens of structure I or III, or either benzylic hydrogen of structure II does not generate isomers, but rather, identical compounds.

11.2 Enantiotopic Groups and Faces

When the application of the substitution criterion leads to nonidentical structures, the groups examined are no longer identical, but rather stereoheterotopic (enantiotopic or diastereotopic) [1-3]. Stereoheterotopic groups are exemplified by the two constitutionally similar substituents a in structure IV.

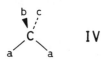   IV

Let us apply the three previously defined criteria (Section 11.1) to structure IV, where the four groups a , a , b , c are taken as achiral. No simple axis of symmetry exists here, and the two stereoheterotopic groups can be interchanged only by a rotation-reflection operation (S_n) to provide a structure indistinguishable from the original. In turn, replacement of one of the two groups, a, with an achiral group, a' (a' ≠ a), generates enantiomers. Finally, the molecular environments of the two groups are enantiomeric; the a groups on the left-hand and right-hand side of struc-ture IV "see" the groups a , b , and c , in a counterclockwise and clockwise sequence, respectively. In such cases, the stereoheterotopic groups are specified as being enantiotopic. A well-known example is that of the methylene hydrogens of ethanol (Va); their monosubstitution with a deuterium atom generates enantiomers. Thus, replacement of H_1 with D yields (S)-ethanol-1-d (Vb), while replacement of H_2 yields (R)-ethanol-1-d (Vc).

A center bearing enantiotopic groups is obviously achiral, but, never-theless it is quite different from a center carrying equivalent groups.

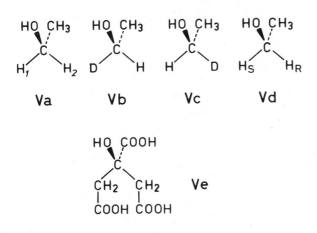

Over a decade ago, the concept of <u>prochirality</u> was introduced by Hanson [4]; thus a center carrying enantiotopic groups is characterized as being <u>prochiral</u> (e.g., IV, V).

Biochemists were the first to recognize the nonidentity of the two groups <u>a</u> in a structure such as IV. During the 1940s, it was demonstrated that the $-CH_2COOH$ groups in citric acid (Ve) are of different biological origins, and that they are degradated differently; thus, these two groups are discriminated by the enzymes involved in their anabolism and catabolism [5]. A proposal was made in 1954 to call a carbon atom of the type C(aabc) a "<u>meso</u>-carbon atom" [5]. This proposal has not gained unrestricted acceptance, as opposed to the above-mentioned and more general concept of prochirality [4].

With the increasing recognition of prochiral centers, it became necessary to clearly label and discriminate between enantiotopic groups, and an extension of the sequence rule procedure made this labeling process feasible [4]. If, in structure Va, H_1 is arbitrarily preferred over H_2 the sequence rule can be applied, resulting in a (S)-configuration, while the preference of H_2 over H_1 leads to the (R)-configuration. As a consequence, H_1 and H_2 are designated <u>pro</u>-S and <u>pro</u>-R, respectively, and are represented as H_S and H_R (Vd) [4]. The prefix <u>pro</u> emphasizes that the groups labeled in this manner belong to a prochiral center. Further proposals decide the more complex cases [2,6].

The concept of prochirality can also be applied to trigonal centers, i.e., to faces of suitable molecules. In acetaldehyde (VI), the two faces of the molecule are not equivalent, but are enantiotopic. Upon viewing the molecule from above, the ligands define a clockwise path, while a view from below affords a counterclockwise path. The upper face is

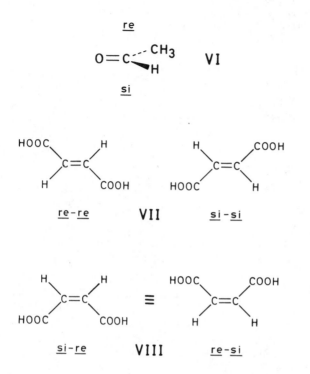

designated re-face (from rectus) and the lower face is called the si-face (from sinister) [4]. In the case of a C-C double bond, the re-si nomenclature is used for both ends independently. Fumaric acid (VII) thus has a re-re- and a si-si-face, while maleic acid (VIII) has two equivalent faces (re-si) due to the presence of a C_2 axis [3].

11.3 Diastereotopic Groups and Faces

Diastereotopic groups are differentiated from enantiotopic groups by the same three criteria used previously. Diastereotopic groups reside in diastereoisomeric environments, cannot be interchanged by any symmetry operation, and upon substitution by chiral or achiral groups lead to diastereoisomeric structures.

The presence of enantiotopic groups in a molecule necessarily implies an element (e.g., a center) of prochirality, while diastereotopic groups imply prostereoisomerism either as an element of prochirality, or of proachirality. In chloroethylene (IX), the two geminal hydrogen atoms are

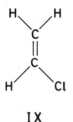

IX

diastereotopic; replacement of one hydrogen affords (Z)- and (E)-diastereoisomers. But no element of prochirality exists in structure IX, and the carbon atom carrying the two hydrogens is designated as a pro-achiral center.

In 1-phenylisopropanol (X), C-2 is an asymmetric carbon, and the molecule is chiral. Carbon-1, though, is prochiral; replacement of one of the stereoheterotopic hydrogens by a group different from the other ligands would generate diastereoisomeric molecules, due to the presence of two centers of chirality. Therefore, the two hydrogen atoms at C-1 are dia-stereotopic groups adjacent to a prochiral center.

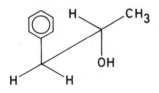

X

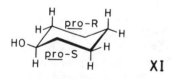

XI

Cyclohexanol (XI) combines the concepts of prochirality and pro-achirality, thus affording an example more complex than the previous ones. Cyclohexanol is achiral, but contains five centers of prochirality (carbons 1, 2, 3, 5, and 6). The two stereoheterotopic ligands of carbon-1 are the two edges of the ring; these two edges are enantiotopic, and are referred to as pro-R and pro-S. The four other prochiral carbons carry diastereo-topic hydrogens. But carbon-4 carries two diastereotopic hydrogen atoms, and is itself proachiral (replacement of either hydrogen yields achiral dia-stereoisomeric 1,4-disubstituted cyclohexane derivatives).

TABLE 1   Relationships Among Constitutionally Similar Groups
in Molecules[a]

Type of groups	Symmetry criterion	Substitution criterion	Elements of prostereo-isomerism
Homotopic (equivalent)	Interchangeable by C_n $(\infty > n > 1)$	Substitution with test groups yields no isomers	None
Enantiotopic	Interchangeable by S_n only	Substitution with achiral or chiral test groups yields enantiomers or diastereoisomers, respectively	e.g., Prochiral center
Diastereotopic	Not interchangeable by any symmetry operation	Substitution with chiral or achiral test groups yields diastereoisomers	e.g., Prochiral or pro-achiral center

[a]Modified from Mislow and Raban [1].

Diastereotopic faces also exist.   While cyclohexanone itself has two equivalent faces, its monosubstituted derivatives have two diastereotopic faces.   This is exemplified by the achiral 4-methylcyclohexanone (XII), and by the chiral 3-methyl- and 2-methylcyclohexanone (XIII).

The properties and conclusions discussed in Chapter 11 are summarized in Table 1; such a simplified overview was originally presented by Mislow and Raban [1] to assist with the diffusion of their newly crystallized rules.   The concepts of topic relationships and of prostereo-isomerism soon proved to be of considerable significance in understanding and rationalizing spectroscopic results, as well as enzymic and stereochemical reactions.   These topics constitute part of Chapter 12.

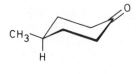

XII

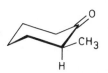

XIII

References

1. K. Mislow and M. Raban, Stereoisomeric relationships of groups in molecules, in Topics in Stereochemistry, vol. 1 (edited by N. L. Allinger and E. L. Eliel), Wiley, New York, 1967, pp. 1-38.

2. H. Hirschmann and K. R. Hanson, J. Org. Chem., 36:3293 (1971).

3. D. Arigoni and E. L. Eliel, Chirality due to the presence of hydrogen isotopes at noncyclic positions, in Topics in Stereochemistry, vol. 4 (edited by E. L. Eliel and N. L. Allinger), Wiley, New York, 1969, pp. 127-243.

4. K. R. Hanson, J. Am. Chem. Soc., 88:2731 (1966).

5. P. Schwartz and H. E. Carter, Proc. Natl. Acad. Sci. USA, 40:499 (1954).

6. H. Hirschmann and K. R. Hanson, Tetrahedron, 30:3649 (1974).

12

Principles of Stereochemical Methodology

"Stereochemical methodology" means the study of all methods, experimental or theoretical, chemical or physicochemical, static or dynamic, analytical or synthetic, which give explicit consideration to the stereochemical aspects inherent in any chemical system. A detailed presentation of stereochemical methods would require several volumes and is far beyond the scope of the present book. Therefore, the following pages will be restricted to a discussion of the general principles in stereochemical methodology, and to a survey of the main techniques.

12.1 Theoretical Principles Underlying the Discrimination of Enantiomers and that of Diastereoisomers

The sharp distinction made in Section 4.1 between enantiomers and diastereoisomers may have seemed somewhat academic. In stereochemical methodology, however, this distinction acquires its real significance.

In Chapter 5, the various nomenclatures used to define the absolute configuration of enantiomers are described. The careful reader will have noticed that all these nomenclatures make use of an external convention on which the definition of absolute configuration is based (e.g., Fischer projection in the D and L nomenclature; clockwise and counterclockwise rotation in the R and S nomenclature). Inverting the convention would result in an inverted assignment.

This discussion can be generalized by considering a chiral molecule such as Cabcd (I). When drawing the enantiomer in the three-dimensional reference frame of right-handed Cartesian coordinates (II) and in its enantiomeric system of axes (III), we obtain the two enantiomeric representations of Cabcd. Indeed, the coordinates of the ligands c and d are,

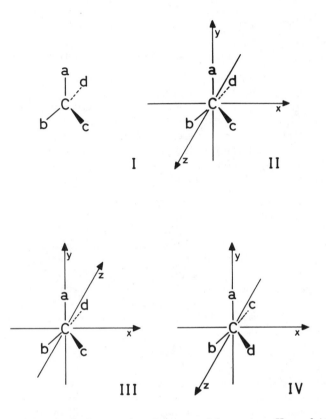

I

II

III

IV

respectively (+, -, +) and (+, -, -) in system II, and (+, -, -) and
(+, -, +) in system III. The ligands c and d have inverted coordinates in
these two reference frames, while the ligands a and b remain unaffected:
(o, +, o) and (-, -, o), respectively. On the other hand, the drawings II
and IV show enantiomeric molecules in the same right-handed system of
axes. It follows that III and IV represent the same enantiomer of Cabcd,
but drawn in enantiomeric reference frames (identical coordinates of all
atoms in the molecule).

Let us allow outselves a short digression by considering the two
enantiomeric reference frames used above. While their absolute configuration is usually accepted without comment, it must be realized that these
reference frames call for their own chiral reference, which is nothing less
than the universe we inhabit. Indeed, our world is chiral, and its asymmetry appears as an intrinsic property of matter itself [1]. The concept
of dissymmetric worlds is a fascinating one, and was remarkably well-
presented a few years ago by Abernethy [2]. In our example, the two

reference frames II and III symbolize two enantiomeric universes containing the object I. Their configuration is defined with reference to the universe (our own) containing them, and as such, is termed "absolute."

A close analogy exists between stereochemical nomenclatures and
stereochemical methodology. The previous paragraphs have shown that
the chirality of a molecule (or any chiral object) can be defined only with
respect to a chiral reference. Similarly, the experimental discrimination
or characterization of enantiomers necessitates the obligatory use of a
"chiral handle." Chiral (dissymmetric) handles include dissymmetric
reagents or solvents, enzymes, and enantiomeric circularly polarized
components of plane-polarized light [3].

As opposed to enantiomers, the definition of diastereoisomers does
not need any external reference. As stated several times throughout the
previous chapters, diastereoisomeric molecules differ in interatomic distances (some atoms or groups of atoms have different spatial relationships). For example, a table of all intramolecular distances suffices to
fully characterize one molecule with respect to all its diastereoisomeric
forms. A table of interatomic distances is called a distance matrix and
does not include any reference other than the molecule itself, as opposed
to a set of three-dimensional atomic coordinates. Examples of distance
matrices can be found in Section 4.3.

In summary, a consequence of differences in topology is that the absolute configuration of achiral diastereoisomers is defined without recourse
to a reference frame. Another consequence is that the energy contents of
diastereoisomers is different, and that all their properties (chemical,
physical, and physicochemical) will be different, however small the difference. Their discrimination by any experimental method is thus conceptually feasible without recourse to a stereochemical handle.

The straightforward discrimination of diastereoisomers is another way
to explain the necessary role of a dissymmetric handle (see above) in distinguishing enantiomers. Take two enantiomeric compounds (A and $\overline{A}$)
interacting with a chiral handle B; the two resulting interactions A-B and
$\overline{A}$-B are stereoisomeric, and more precisely diastereoisomeric. As a
consequence, any suitable symmetrical technique will discriminate between
A-B and $\overline{A}$-B.

The above statements (chiral handle is obligatory for the discrimination of enantiomers, but unnecessary for the discrimination of diastereoisomers) are, of course, correct. This Artistotelian-like dichotomy is,
however, misleading since it does not make clear diastereoisomeric
interactions between enantiomers (i.e., a given chiral molecule acting as
a dissymmetric handle for its own enantiomer). Two enantiomers
examined separately by any technique will show identical behaviors under

identical conditions (excluding a chiral handle). The progressive contamination of a chiral compound with its enantiomer may lead to the alteration of some characteristic properties in the solid state (e.g., melting point) or in the dissolved state (e.g., optical rotation, chemical shifts in nuclear magnetic resonance) (see, for example, Refs. 4 and 5). These alterations may be due to the simultaneous presence of two diastereoisomeric complexes, in the case of dimers, A-A and A-Ā.

12.2 Classification of Stereochemical Techniques

The techniques used in stereochemical methodology can be characterized according to three main goals and can be classified accordingly. The resulting classes of techniques are represented in Figure 1 as three overlapping circles.

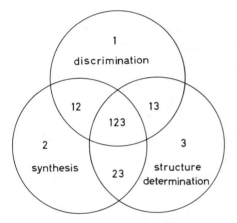

FIGURE 1 A general classification of stereochemical techniques.

The three main groups of techniques are:

Circle 1: the discrimination of stereoisomers. The term "discrimination" must be understood as meaning either

  The physical separation of stereoisomers, which is accomplished by
      such techniques as the resolution of enantiomers via diastereo-
      isomeric salt formation, or the chromatographic separation of
      diastereoisomers;
  The discrimination of spectroscopic signals (e.g., in nuclear magnetic
      resonance) generated by stereoisomers.

Circle 2: the synthesis and chemical transformation of stereoisomers. To this group belong all synthetic techniques involving stereoisomers as reagents and/or as products of reaction (e.g., asymmetric synthesis, derivatization of chiral molecules with chiral reagents).

Circle 3: the structure determination of stereoisomers. This field covers all methods allowing assessment of

The absolute three-dimensional configuration of chiral molecules and diastereoisomers
The conformation of molecules, together with thermodynamic factors, opening the door of the temporal dimension.

The overlapping parts of circles 1-3 have well defined meanings. They symbolize experimental procedures combining the goals or techniques of two or three circles.

The overlap of circles 1 and 2 (Field 12) represents these synthetic procedures used in the separation of enantiomers, for example, the synthesis of diastereoisomeric derivatives suitable for separation by crystallization or chromatography.

Field 13 represents those methods used to correlate the configuration of stereoisomers with the molecular properties involved in their discrimination (e.g., assessment of configuration based on chromatographic behavior or spectroscopic signals).

Field 23 represents structure determination by unambiguous synthetic routes (synthetic stereochemical correlations).

Field 123 symbolizes such procedures as structural correlations based on discrimination of diastereoisomeric derivatives of enantiomers.

More details will be needed for a good understanding of the above classification. The following sections cover the main aspects of stereochemical methodology, using this classification as a reference in order to better characterize the methods discussed and their goals.

12.3 Optical Activity

Plane-polarized light as generated in a polarimeter can be considered as being the resultant of a left-handed and a right-handed circularly polarized in-phase beam. Thus, vector E of the electrical field associated with the beam is the result at any moment of the two vectors E_R and E_L (V). The beam travels along axis $\underline{z}$ so that the two vectors E_R and E_L trace out helical paths of opposite chirality (P and M, respectively). Each of these enantiomeric paths constitutes a "chiral handle" as defined in Section 12.1.

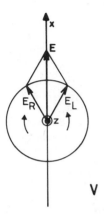

V

When a beam of polarized light travels across a dissymmetric medium, the interactions of the medium with the two circularly polarized components of the beam show a diastereoisomeric relationship [6].

A dissymmetric medium is thus capable of discriminating between the two enantiomeric components of plane-polarized light. This discrimination is experimentally detected as differences in the refractive indices of the two components ($n_R \neq n_L$) and as differences in the molar extinction coefficients ($\epsilon_R \neq \epsilon_L$). The difference in refractive indices n_R and n_L is called circular birefringence and corresponds to the slowing of one circular component relatively to the other. As a result, the plane of polarization is rotated (VI) by an angle α called the angle of rotation, and this phenomenon is called the optical rotation. A medium showing optical rotation is called optically active.

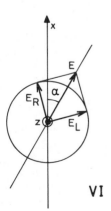

VI

The angle of rotation α in degrees per centimeter is given by Eq. (1):

$$\alpha = \frac{180}{\lambda} \, (n_L - n_R) \tag{1}$$

The magnitude of α for a given solute is dependent on wavelength (λ), solvent, temperature (T), and also on the number of molecules in the path of light, assuming no interaction between these molecules which would result in a deviation for ideality. For a discussion of nonideality and specific examples, see Refs. 4 and 6a.

Results of optical rotation are expressed in <u>specific rotation</u> $[\alpha]$ or in <u>molecular rotation</u> $[\phi]$ at a given λ and T:

$$[\alpha]_\lambda^T = \frac{\alpha_\lambda^T \cdot 100}{L \cdot c} \tag{2}$$

$$[\phi]_\lambda^T = \frac{MW \cdot [\alpha]_\lambda^T}{100} \tag{3}$$

where α is in degrees, L is the path length in dm, and c is the concentration of the solute in g/100 ml. In a mixture of enantiomers, the <u>optical purity</u> is defined by the relation:

$$\% \text{ optical purity } = \frac{[\alpha]}{[\alpha]_o} \cdot 100 \tag{4}$$

where $[\alpha]$ is the specific rotation of the enantiomeric mixture and $[\alpha]_o$ is the specific rotation of one pure enantiomer. Assuming ideal behavior (no enantiomeric interactions and obedience to the Beer-Lambert law) the optical purity is equal to the <u>enantiomeric purity</u>; for R > S:

$$\% \text{ enantiomeric purity } = \frac{(R - S)}{(R + S)} \cdot 100 \tag{5}$$

where R and S are the concentrations of the two enantiomers [6,7].

Finally, the <u>enantiomeric percentages</u> (or enantiomeric compositions) are derived from the relations:

$$\% \text{ (R)-enantiomer } = \frac{R}{(R + S)} \cdot 100 \tag{6}$$

$$\% \text{ (S)-enantiomer} = \frac{S}{(R + S)} \cdot 100 \tag{7}$$

Enantiomeric percentages can be calculated from the enantiomeric purity, e.g.,

$$\frac{R}{(R + S)} \cdot 100 = \frac{(R - S)}{(R + S)} \cdot 100 + \frac{100 - \frac{(R - S)}{(R + S)} \cdot 100}{2} \tag{8}$$

Measurements of optical activity are conventionally carried out at the wavelength of the sodium D-line (589 nm), but spectropolarimeters render accessible the spectral region scanned by ultraviolet and visible spectrophotometers. The dependence of molecular rotation on wavelength is termed optical rotatory dispersion (ORD) and will be referred to again in Section 12.6.

The difference in molar extinction coefficients results from a phenomenon known as circular dichroism (CD). Since one of the circularly polarized components of the plane-polarized beam is absorbed more than the other, the resultant vector E no longer oscillates in a plane (appearing as a line in projections V and VI) but traces out a flattened helix (whose projection in VII is an ellipse). The difference $\epsilon_L - \epsilon_R$ is designated $\Delta\epsilon$, the differential absorptivity, and is also expressed by the molar ellipticity (or molecular ellipticity) [6]:

$$[\theta]_\lambda^T = 3300 \Delta\epsilon \tag{9}$$

VII

The molar ellipticity is measured as a function of wavelength using an instrument called a dichrograph. The wavelength-dependence of circular birefringence and circular dichroism constitutes the Cotton effect whose stereochemical applications will be discussed briefly in Section 12.6

Let us now turn our attention to the molecular features responsible for optical activity. An extremely clear description of the relation between rotation and configuration has been given by Eliel [8]. A sufficient (but not necessary) condition (see below) for optical activity is the presence of an asymmetric center in the molecule. For example, the configuration shown in VIII is dextrorotatory when the polarizabilities of the ligands are in the order $R_1 > R_2 > R_3 > R_4$. The polarizability of selected ligands is in the order

$$I > Br > SH > C{\equiv}C > C{=}C \equiv Phenyl > Cl > CN > COOH > CH_3 > NH_2 > OH > H > D > F$$

It follows, for example, that (S)-hydratropic acid (VIII; R_1 = phenyl, R_2 = COOH, R_3 = CH_3, R_4 = H) is dextrorotatory. It must be noted, however, that the differences in polarizabilities of ligands may be so minute as to result in an undetectable optical activity; the compound is de facto optically inactive. Such is the case of n-butyl-ethyl-n-hexyl-n-propyl-methane (VIII; R_1 = C_6H_{13}, R_2 = C_2H_5, R_3 = C_4H_9, R_4 = C_3H_7) [9,9a]. The reader interested in the basic principles, theories, and models of optical activity is referred to an excellent review by Brewster [10].

VIII

A chiral center in a molecule is not a necessary condition of optical activity because the symmetry of the whole molecule must be considered. The necessary and sufficient condition is for a molecule to be chiral (no reflection symmetry, see Section 1.2). Any element of chirality (center, axis, plane, helicity) in a molecule will render it optically active, even if the optical rotation is vanishingly small with respect to the current technology [11]. Such a statement is obvious to the reader familiar with the previous chapters; it was, however, not that obvious a few decades ago, when the elements of chirality had not been clearly characterized.

The information obtained from optical rotation data is manifold. The simple sign of rotation, (+) or (-), is extensively used to label enantiomers. While the wavelength of the sodium D-line is assumed, there is nothing in such a label to specify the solvent. This point may prove critical in some cases, since several molecules are known to exhibit a reversal of rotation when examined in solvents of different polarities, a phenomenon accounted for by solvent-induced conformational changes. As implicit in Eq. (4) specific rotation affords a straightforward and popular method for determining the enantiomeric purity of a chiral compound, or the enantiomeric composition of a mixture of enantiomers. As such, these uses of optical rotation resort to discrimination of stereoisomers (Figure 1, circle 1).

Optical rotation can also be studied as a function of time. As a result, such stereochemical reactions as mutarotation and racemization can be investigated, and their kinetics assessed. The topic of optically labile compounds and their study by polarimetry has been reviewed [12].

Finally, measurements of optical rotation at a single wavelength may be used to determine absolute configurations (Figure 1, circle 3), mainly by the method of correlation of configurations (for a review, see Ref. 13). Application of the polarizability rule (see above) allows direct correlation of absolute configuration and rotation in simple cases. Such assignments, however, are not always beyond doubt, and must be reinforced by the comparison with rotations of closely related compounds of known absolute geometry.*

Of wider application is the Displacement Rule formulated by Freudenberg [14]. This rule states that two molecules of similar constitution and identical chirality, when chemically altered in the same way, will display changes in molecular rotation in the same direction and usually of the same order of magnitude. For reliable correlations to be made, similarity of structure is necessary, and an extended series of comparisons is desirable.

12.4 Stereochemical Reactions; Stereosynthesis

Stereochemical reactions constitute a field of utmost significance in the study of both stereochemistry and synthetic chemistry. The main and most obvious goal of stereosynthesis is the preparation of specific stereoisomers (Figure 1, circle 2), for whichever purpose these may be required. Other significant goals should, however, not be forgotten; they include

---

*But the best reinforcement usually comes from ORD and CD curves, discussed in Section 12.6.

stereochemical correlations by chemical synthesis (Figure 1, field 23), and the stereochemical derivatization of enantiomers for analytical purposes (Figure 1, field 12).

The terms "stereoselectivity" and "stereospecificity" refer to stereo-synthesis; their meaning, unfortunately, remains subject to various interpretations aptly summarized by Eliel [15]. This authority calls "stereoselective synthesis" a synthesis producing one diastereoisomer (or a diastereoisomeric (±)-pair) preferentially over all other possible dia-stereoisomers. Stereoselectivity in this meaning does not usually concern the synthesis of optically active compounds. The term stereoselective can be modulated either semiquantitatively (e.g., highly stereoselective) or quantitatively (e.g., 90% stereoselective). On the other hand, Eliel uses "stereospecific" to qualify a reaction where stereoisomeric starting mate-rials (substrates) give rise to stereoisomerically different products. Using this definition, all stereospecific processes are stereoselective, but not all stereoselective processes are stereospecific.

In accordance with trends in biochemistry and chemical pharmacology [16] we favor another definition of stereoselectivity. Any chemical process is termed stereoselective when one stereoisomer is preferred over the other(s); the definition applies to enantiomers and diastereoisomers alike. When a reaction yields stereoisomeric products, and when one of them pre-dominates over the others, the designation of product stereoselectivity is used. When, on the other hand, two or more stereoisomeric substrates are consumed at different rates in a given reaction, the latter is said to display substrate stereoselectivity. A combination of both selectivities has been characterized in biochemistry, namely, when two isomeric substrates (e.g., S_1 and S_2) generate two isomeric products (e.g., P_1 and P_2) in dif-ferent ratios (e.g., S_1 giving rise to more P_1 than P_2, and S_2 to more P_2 than P_1). The term substrate-product stereoselectivity has been used to label such a situation [16].

Stereochemical reactions are based on thermodynamic principles just like any chemical reaction. Let us consider the general case of a molecule, M (the substrate), being transformed into the enantiomeric products, P_R and P_S, by a given chemical reaction (IX). If the two transition states lead-ing to P_R and P_S are themselves enantiomeric (Figure 2), they will have identical energy contents (identical free energy of activation $\Delta G^{\ddagger}$). Applica-tion of Eyring equation [Eq. (1), Section 4.1] shows that the rate constants

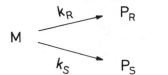

IX

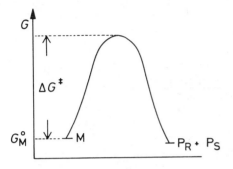

FIGURE 2 Energy profile of two competitive reactions involving enantiomeric transition states and enantiomeric products (IX, $k_R = k_S$).

k_R and k_S will be identical, and that the reaction will yield the products P_R and P_S in identical amounts.

When the transition states are diastereoisomeric (Figure 3), the free energies of activation $\Delta G_R^{\ddagger}$ and $\Delta G_S^{\ddagger}$ must differ ($\Delta\Delta G_{R-S}^{\ddagger} \neq 0$), as differ the two rate constants k_R and k_S ($\Delta k_{R-S} \neq 0$). Further, if the reaction $M \rightarrow P$ is an <u>irreversible</u> one, it can be deduced from Figure 3 that P_R and P_S will be produced in different yields ($P_S > P_R$ in our example). The ratio $P_R{:}P_S$ is a function of $\Delta\Delta G^{\ddagger}$ just as the relative populations of two conformers are a function of their energy difference ΔG° (Section 9.1). Figure 1 (Section 9.1) can also be used to correlate the product ratio $P_R{:}P_S$ of an irreversible stereochemical reaction to the difference of the

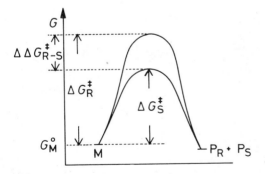

FIGURE 3 Energy profile of two irreversible competitive reactions involving diastereoisomeric transition states and enantiomeric products (IX, $k_R \neq k_S$).

free energies of activation of the diastereoisomeric transition states. At
T = 300°K, a $\Delta\Delta G^{\ddagger}$ of 11.5 kJ/mol (2.75 kcal/mol) results in a 100:1
product ratio.

An irreversible reaction with diastereoisomeric transition states as
shown in IX and Figure 3 is said to be <u>kinetically controlled</u> ($k_R \neq k_S$). The
same terminology applies if the products of a stereochemical irreversible
reaction are diastereoisomers, e.g., P_E and P_Z (X), rather than enantiomers (IX). The ratio $P_E:P_Z$ in this case is dependent upon the rate constants k_E and k_Z, i.e., upon $\Delta\Delta G^{\ddagger}_{Z-E}$ (Figure 4).

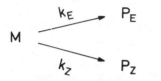

Should the stereochemical reaction be reversible (XI), an equilibrium
establishes itself between the substrate M and the diastereoisomeric
products P_E and P_Z. The equilibrium ratio $P_E:P_Z$ in this case is
<u>independent</u> of k_E and k_Z, but will be a function of the sole difference in
the ground state free energies of P_E and P_Z (ΔG^{o}_{Z-E}). Such a reversible
stereochemical reaction is said to be <u>thermodynamically controlled</u>.

A question of interest in stereosynthesis is whether the <u>conformational
state of the substrate</u> may influence the reaction. Indeed, the preferential
reactivity of one conformer of a flexible substrate M would be a case of
substrate stereoselectivity. The <u>Curtin-Hammett principle</u> [17] explains

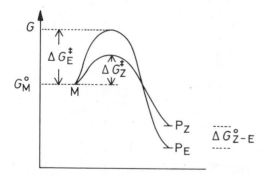

FIGURE 4 Energy profile of two competitive reactions involving diastereoisomeric transition states and products; irreversible reactions (X),
kinetic control; reversible reactions (XI), thermodynamic control.

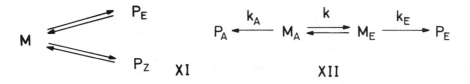

XI XII

how the preferred conformation of a substrate must not necessarily be the reactive conformation. When the barrier of conformational conversion is much smaller than the activation energy of the reaction, the preferred conformation has no relevance to, and no predictive value for, the nature of the product(s). This is illustrated in XII and Figure 5 for a compound M existing in the two conformations M_A and M_E and yielding the corresponding products P_A and P_E. Under irreversible conditions, the product P_E generated by the path of lower energy will be produced in much larger amounts than the alternate product P_A. This selectivity is independent from the relative populations of M_A and M_E; in our example, the less abundant conformer is selected by the reaction. A general equation of relationship between products' ratio and conformational equilibrium has been formulated recently [17a]; one of the limiting cases covered by this general equation relates to the Curtin-Hammett principle.

After having laid down some principles underlying stereochemical reactions, it is useful to review briefly some of these reactions. Table 1 lists the types of stereochemical reactions to be considered in the following pages, together with representative examples. The latter are aimed at illustrating the general principles of stereochemical reactions and facilitating their understanding.

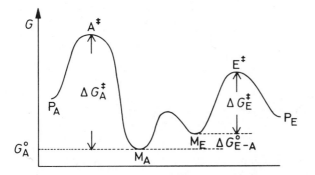

FIGURE 5 Energy profile of a hypothetical reaction XII where the conformational behavior of the substrate M does not influence the product selectivity.

TABLE 1 Important Types of Stereochemical Reactions Discussed in Text

---

Addition reactions at olefinic groups

Electrophilic trans-addition (XIII, XIV)
cis-Hydroxylation (XV)
Diels-Alder reaction (XVI)

Elimination reactions at diastereoisomeric molecules

Substrate stereoselective trans-E_2 elimination (XVII)

Substitution reactions at chiral centers

Racemization: SN_1 (XVIII)
Inversion of configuration: SN_2 (XIX)
Retention of configuration: SN_i (XX)

Elimination reactions at chiral centers

Substrate stereoselective $sp^3 \rightarrow sp^2$ conversion, with chiral
reagents (XXVII)

Reactions at enantiotopic groups and faces

Esterification of enantiotopic acyl groups by a chiral alcohol
(XXII)
Stereoselective reduction of prochiral ketones by chiral
reagents (XXIII)

Kinetic resolution of rapidly interconverting enantiomers

Formation of optically active peptides from racemic
5-oxazolones and (S)-amino acid esters (XXIV)

Reactions at diastereotopic groups and faces

Stereoselective reduction of a prochiral keto group in chiral
ketones by achiral reagents (XXV, XXVI)

Enzymic reactions

Substrate stereoselective oxidation of alcohols catalyzed by
alcohol:NAD oxido-reductase (XXVII)
Substrate and product stereoselective reduction of ketones by
the same enzyme.

Addition reactions at olefinic groups may conceivably result in a mixture of diastereoisomeric products. In fact, the mechanisms of reaction usually have strict steric requirements, and high product stereo-selectivity is the rule. Consider, for example, the reactions of electro-philic addition to carbon-carbon double bonds; the two-step mechanism imposes that the second substituent enters opposite from the first sub-stituent, resulting in a trans-product (XIII). Even in acyclic molecules a single diastereoisomer is formed, such as the threo-product in example XIV [18].

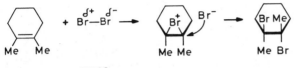

XIII

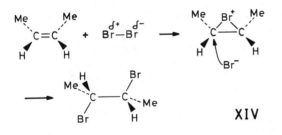

XIV

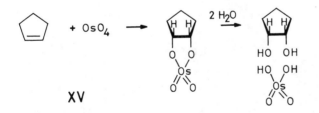

XV

Other addition reactions at alkenes result in cis-products, as, for example, the cis-hydroxylation mediated by osmium tetroxide (XV) or potassium permanganate. In this case, the cis-configuration is imposed because the hydrolytic step cleaves the Os-O bonds and not the C-O bonds [18].

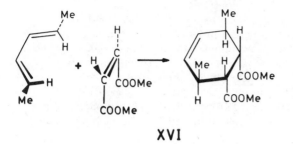

XVI

The Diels-Alder reaction involves the addition of a dienophile (e.g., XVI, methyl maleate) to a diene (e.g., trans-trans-1,4-dimethylbutadiene). The dienophile approaches the diene with an endo-relative position, and the cycloaddition occurs with stereoselective cis-configuration (XVI). Many other stereoselective olefinic additions are known [18, 19].

The reaction of E_2 elimination (at single or double C-C bonds) involves a base-catalyzed removal of a proton (or another cationic group, e.g., Br^+) from the β carbon, with simultaneous loss of a leaving group (halide ion, tosylate, etc.) from the α carbon (e.g., XVII). The electron pair released by the proton abstraction is shifted to the α carbon on the side opposite to the halogen; the favorable configuration of the two substituents eliminated by E_2 is thus trans. The reaction shows a high degree of substrate stereoselectivity in that cis elimination is slower by several orders of magnitude [20]. As a whole, all reactions discussed so far show stereoselectivity because the alternate, diastereoisomeric transition states are highly improbable for steric reasons.

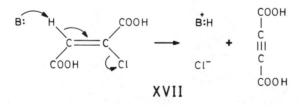

XVII

Reactions at a chiral center may proceed with racemization, inversion of configuration, retention of configuration, or with the loss of chirality. The first three types result from substitution reactions; these are the well-known SN_1 reactions with a planar carbonium transition state resulting in racemization (XVIII), the SN_2 reactions which invert the configuration through a pentacoordinate transition state (XIX), and the SN_i reactions accompanied by retention of configuration (XX) [e.g., 21,22].

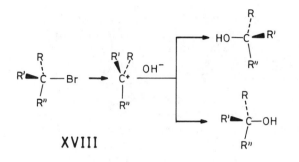

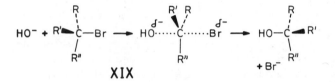

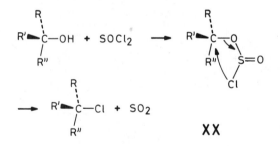

Reactions of elimination at chiral centers will result in the loss of chirality ($sp^3 \rightarrow sp^2$). If the reagent itself is also chiral, each enantiomeric substrate will yield a diastereoisomeric transition state. As a result, the reaction will exhibit substrate stereoselectivity. An example of such a reaction is the dehydrogenation of a chiral secondary alcoholic group to an achiral keto group; since the reaction is not uncommon in biochemistry, a relevant example will be discussed later when considering the stereochemistry of enzymic processes.

Asymmetric synthesis is properly defined as comprising those reactions which create an element of chirality in substrate molecules, and which occur with product stereoselectivity. In other terms, an asymmetric synthesis converts a prochiral unit into a chiral unit with resulting unequal amounts of stereoisomeric products [23].

The first condition of an asymmetric synthesis is the presence of a prochiral unit in the substrate molecule, be it enantiotopic or diastereotopic groups or faces. Beginning with a simple example, let us consider the reduction of a prochiral ketone with lithium aluminum hydride (XXI); attack of the achiral reagent at either enantiotopic face yields two enantiomeric transition states of identical energy and probability, and the product of the reaction is a racemic mixture of the two enantiomeric alcohols. Such a reaction does not pertain to asymmetric synthesis because the second condition is not met, namely, asymmetric induction. The latter is brought forth by an element of chirality playing an active role in the reaction: chiral reagent, chiral solvent, chiral catalyst, circularly polarized light, or element of chirality in the substrate molecule [23]. By "active role" it is meant that the element of chirality is part and parcel of the transition states, which, as a result, are diastereoisomeric. A far-reaching theory of diastereoisomeric transition states and asymmetric induction has been developed recently by Salem [23a].

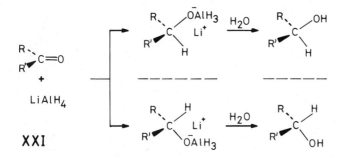

Countless examples are known of stereoselective reactions at enantiotopic groups or faces by a chiral reagent [e.g., 23]. One such example is detailed in scheme XXII, namely the reaction of β-phenylglutaric anhydride with (-)-menthol. The nucleophilic attack of the alcohol can occur either on the carbonyl carbon of the pro-R branch, or on that of the pro-S branch. The two products of the reaction are the diastereoisomeric monomenthyl esters; in the condition of the study, the stereoselectivity of the reaction was very low (54:46 ratio), possibly due to a thermodynamic rather than kinetic control [24].

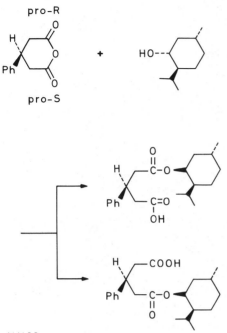

XXII

An example of asymmetric synthesis involving an enantiotopic face is seen in the reduction of isopropyl phenyl ketone by the Grignard reagent from (+)-1-chloro-2-phenylbutane (XXIII). Depending upon the face attacked by the reagent, two diastereoisomeric transition states are generated, and the reaction shows a significant product stereoselectivity [25].

As defined above, asymmetric synthesis implies the creation of an element of chirality. The example to be presented now is relevant to the present context despite the fact that it does not strictly qualify as an asymmetric synthesis. Racemic 5-oxazolones such as 4-isopropyl-2-phenyloxazolone react with (S)-amino acid esters to give optically active peptides (XXIV), more than 50% of the racemic-starting material being incorporated into the peptide as one enantiomer [25a]. The highest selectivity was obtained when the amino acid was valine (R = isopropyl); the (R;S):(S;S) ratio of diastereoisomeric products was 83:17 rather than 50:50. This was rendered possible by the base-catalyzed racemization of the oxazolones, the latter reaction being faster than the aminolysis reaction. The overall reaction is therefore an example of a kinetic resolution of rapidly interconverting enantiomers.

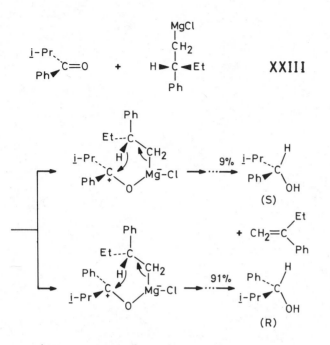

Asymmetric synthesis by <u>asymmetric catalysis</u> is an actively investigated field of great promises and future. In such reactions, a prochiral element reacts with an achiral reagent under the influence of a chiral catalyst. The reaction may be homogeneous (soluble catalyst) or heterogeneous. Examples are seen in the catalytic hydrogenation of alkenes having enantiotopic faces [23].

Asymmetric synthesis by attack at <u>diastereotopic groups or faces</u> does not require a chiral reagent. Indeed, consider the reduction of a chiral aldehyde as depicted in scheme XXV. Two diastereoisomeric transition states result from the approach of the achiral reagent from opposite directions; a fair stereoselectivity is apparent [26]. Examining the stereoselectivity of addition reactions of chiral aldehydes and ketones allowed Cram to formulate a rule of considerable predictive value. By classifying the substituents of the chiral center according to their size (large, L; medium, M; small, S), Cram showed that the attack of the entering group occurs preferentially from the less hindered side when the largest group is <u>trans</u> to the carbonyl oxygen (XXVI). This generalization is now known as <u>Cram's rule</u> [26]. The model, however, has been continuously refined and extended to accommodate situations involving rigid and dipolar structures [23,26a]. Proposals by Karabatsos [26b] and Felkin [26c,26d] concern the conformation of the transition state. Other generalizations have

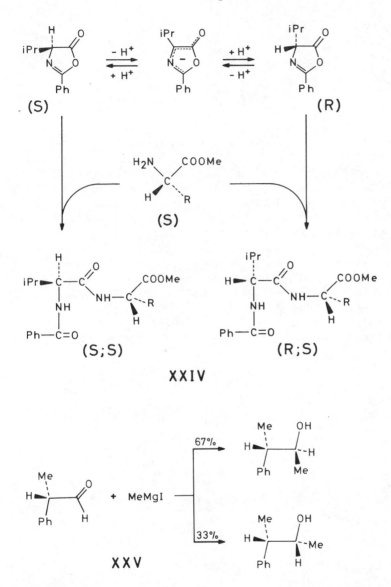

XXIV

XXV

been formulated, for example, Prelog's rule [26e, 26f], and the mathematical model of Ruch and Ugi [26g]. A discussion of these various models is beyond the scope of the present review. The interested reader is referred to the remarkable survey by Morrison and Mosher [23], and to the many references given therein.

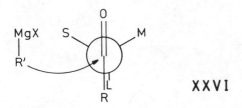

XXVI

    Enzymes are the chiral tools of the chiral biosphere. High stereo-selectivity is the universal rule of enzymic reactions in the transformation of isomeric substrates as well as in the generation of isomeric products [e.g., 27-33a]. This rule is easy to understand when considering the enzymes as chiral reagents, whatever their complexity. As an example, let us examine the metabolism of exogenous 3-methylcyclohexanols by horse liver alcohol dehydrogenase (alcohol:NAD oxido-reductase), an enzyme utilizing nicotinamide adenine dinucleotide (NAD) as the cofactor (see XXVII). The enantiomeric (-)-cis- and (+)-cis-isomers are metab-olized to (+)- and (-)-3-methylcyclohexanone, respectively, with a 1:3 ratio in the reaction rates (XXVII). A marked substrate stereoselectivity is thus apparent with the (+)-enantiomer being a better substrate than the (-)-enantiomer for the sp^3 - sp^2 transformation. On the other hand, the same enzyme catalyzes the reduction of (-)-3-methylcyclohexanone to (+)-cis-

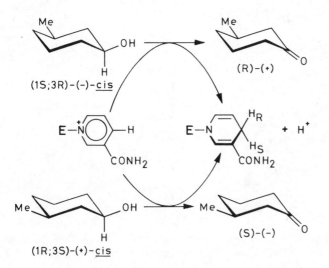

XXVII

and (+)-trans-3-methylcyclohexanol at a rate about 60 times higher than
the reduction of (+)-3-methylcyclohexanone to (-)-cis- and (-)-trans-3-
methylcyclohexanol [34]. This complex example shows that both the
oxidation and reduction catalyzed by liver alcohol dehydrogenase are substrate stereoselective. In addition, the latter reaction is product-
stereoselective in that the cis- and trans-alcohols are consistently formed
in a 2:1 ratio [34].

Other applications exist for stereochemical reactions than the synthesis of stereoisomers as discussed above. These other applications are
kinetic resolution and stereochemical correlations.

Resolution involves the physical separation of enantiomers, and will be
broadly presented in Section 12.5. Kinetic resolution is a particular case
of resolution, based on the difference in the reaction rates of a chiral
reagent with enantiomeric substrates (substrate stereoselectivity). If a
racemic compound is allowed to react with an insufficient amount of a
chiral reagent, the two diastereoisomeric products will be formed in a
ratio depending on the ratio of the two rate constants. At the completion
of the reaction, the unreacted substrate will be enriched in the slow-
reacting enantiomer. While kinetic resolution may be used to achieve
partial resolution of enantiomers, it has also been found to be of value in
determining the absolute configuration and the enantiomeric purity of
chiral compounds [34a]. The quantitative interpretation of the results as
well as more elaborate developments of kinetic resolution have been
described in comprehensive reviews [35,36].

Stereochemical correlations involve the assessment of the unknown
absolute configuration of given molecules by deriving it from the known
absolute configurations of other compounds. Besides spectroscopic,
crystallographic, and optical rotation methods, chemical reactions have
proven to be of great value in stereochemical correlations. Only a brief
outline of the latter method will be given below, and the interested reader
is referred to authoritative reviews [13,37-41]. Correlations by synthetic
routes require these to be stereochemically nonambiguous. It is obvious
that this condition is an absolute prerequisite, and that the design of such
experiments requires much expertise. For the investigation of an asymmetric center, safe and reliable correlations are provided by chemical
reactions which do not displace any atom directly bonded to the center.
The chirality of the center remains untouched, as shown by the correlation
of the configuration of the natural (-)-nicotine with that of (S)-proline (XXVIII)
[37]. This example involves degradative steps, but correlations by synthesis may be just as useful.

Correlations involving the making or the cleavage of a bond at a chiral
center are valid only if the steric course of the reaction is known beyond

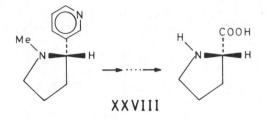

XXVIII

doubts. For such correlations, use is made of mechanisms of unequivocal stereochemistry such as the SN_2 inversion of configuration; kinetic measurements must be made in order to ascertain that the actual mechanism occurring is the postulated one. For example, the replacement of the bromine atom in (S)-(+)-α-bromopropionic acid (XXIX) under kinetically controlled conditions which bring about inversion of configuration allows the correlation of (R)-(-)-alanine (XXX) and (R)-(-)-lactic acid (XXXI) [37].

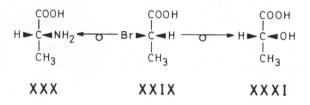

XXX   **XXIX**   **XXXI**

12.5 Physical Discrimination of Stereoisomers

This section discusses the utility of several physical methods in discriminating stereoisomers. The use of melting points will be considered first, followed by the methods of physical separation of stereoisomers, namely, classical resolution and chromatographic techniques.

An assembly of enantiomers in equal amounts, i.e., one-half of the assembly being molecules of the dextrorotatory, the other half molecules of the levorotatory enantiomer, is called a <u>racemic modification</u> and is represented by the symbol (±). In the vapor and in the liquid phase (bulk or solution), the term racemic modification does not need to be specified any further, since these usually behave as ideal or nearly ideal mixtures of enantiomers. Except in infrequent cases of diastereoisomeric interactions of enantiomers in solution (see Section 12.1) [4], racemic modifications in the vapor or liquid phase show physical properties identical to those of pure enantiomers (e.g., boiling point, refractive index, density).

In the solid (crystalline) state, however, the term racemic modification needs clarification. The specific packing forces of the crystalline state are very sensitive to changes in structure, and a chiral molecule will interact differently with an identical molecule or with its enantiomer. As a result, deviations from ideality are the rule in the solid state, and three cases of racemic modifications are encountered. When the solid phase is a mixture of crystals of the dextrorotatory enantiomer, and of crystals of the levorotatory enantiomer (i.e., when the two enantiomers crystallize in separate crystals), the racemic modification is called a racemic mixture (also called conglomerate). The physical properties of a conglomerate are in many respects similar to that of the pure enantiomers. The melting point of a conglomerate, however, is lower than that of the enantiomers, because mixtures of crystals of pure enantiomers are eutectic mixtures. Figure 6 shows a typical melting point diagram (liquidus curve only) for conglomerates.

More common than racemic mixtures are racemic compounds (also called racemates). In this case, an equal number of molecules of each enantiomer associate in the unit cell of the crystal, which will thus contain as many (+)- as (-)-molecules. Racemates melt above, below, or at the same temperature as the enantiomers, as shown by two typical melting point diagrams (Figure 7). Being true compounds, the racemates differ in many physical properties from the enantiomers.

In some rather rare instances, racemic modifications show a close-to-ideal behavior and are called racemic solid solutions. This means that there is little affinity difference in the solid state between molecules of identical or opposite configuration, as opposed to preferential affinity between identical molecules (conglomerates) or preferential affinity between enantiomeric molecules (racemates). Solid solutions have properties identical to that of the enantiomers, including melting points

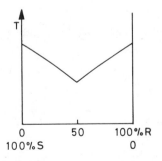

FIGURE 6 Binary melting point diagram of enantiomers forming a racemic mixture.

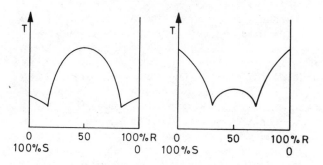

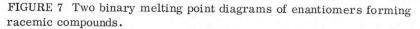

FIGURE 7 Two binary melting point diagrams of enantiomers forming racemic compounds.

(allowing for very slight variations, see Figure 8). An excellent and detailed presentation of racemic modifications has been given by Eliel [42].

Let us now consider the case of two chiral compounds A and B of great structural similarity; (+)-A and (-)-B are called quasienantiomers, as are (-)-A and (+)-B. Melting point diagrams obtained from mixtures of two quasienantiomers may have the general shape displayed in Figure 9a, in which case the 50:50 mixture is called a quasiracemate. In our example, mixtures of (+)-A and (+)-B result in a noncharacteristic melting point diagram (Figure 9b). The determination of melting point diagrams of quasi-enantiomers is a way to determine the absolute configuration of (+)- and (-)-B when that of (+)- and (-)-A is known [37,43].

The physical separation of the enantiomeric components of a racemic modification is called resolution. Together with asymmetric synthesis, it is the most widely used method for the production of optically active

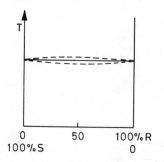

FIGURE 8 Binary melting point diagram of enantiomers forming a racemic solid solution.

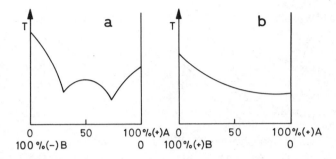

FIGURE 9 An example of binary melting point diagrams obtained with quasienantiomers (Figure 9a); compare with Figure 9b, where the two compounds are not quasienantiomers.

compounds. Resolution is based either upon the physical separation of enantiomeric crystals, or upon the separation of diastereoisomeric derivatives.

The physical sorting of enantiomeric crystals necessitates a racemic mixture. The method has little practical value, but great historical significance. Indeed, the first resolution ever achieved was accomplished in this way by Louis Pasteur in 1848 with the sodium ammonium salt of racemic tartaric acid. Below 27°C, this compound crystallizes in hemihedric crystals, i.e., which have facets forming dissymmetric combinations. The manual separation of such crystals implies that each of them in turn is compared to a standard crystal (which is the chiral handle) and a decision is taken regarding their identity or enantiomeric relationship.

Mechanical separation may also be achieved by the inoculation method, namely, the selective seeding of a solution of racemate with crystals of one enantiomer. An intriguing technique is that of the crystal-picking method [43a]. Individual crystals obtained from solutions of racemic helicenes (see Section 7.3) contain unpredictable proportions of enantiomers; solutions of single crystals (picked up manually) show either optical activity of varying sign and amplitude, or no activity. This suggests that the picked-up crystals are not single conglomerate crystals, but are made up from several of the latter. The method has been of great use in studying the chiroptical properties (see Section 12.6) of helicenes.

By far the most useful method of resolution is by formation of diastereoisomers using a suitable reagent. As already discussed, an optically active compound B reacts with a racemic modification (A + $\overline{A}$) to form the diastereoisomeric derivatives A-B and $\overline{A}$-B. Two main techniques are currently used for the separation of the diastereoisomeric derivatives; these are crystallization and chromatography.

The classical resolution by crystallization is most conveniently achieved by the formation of diastereoisomeric salts. Thus, a racemic acid is reacted with an optically active amine, while a racemic amine is reacted with an optically active acid. The solvent is chosen so that one of the diastereoisomeric salts is markedly less soluble than the other. Salts are popular derivatives because they are easily broken to yield the starting compounds, allowing recovery of the desired product and sometimes also of the reagent. A list of a few common resolving agents is given in Table 2.

The resolution of alcohols may sometimes present some difficulties. Prior to resolution, they are often converted to their acid phthalate or succinate esters; these half-esters are then resolved as typical acids. Alternatively, the racemic alcohol is reacted with an optically active acylating agent (e.g., menthoxyacetyl chloride) to form diastereoisomeric esters. In any case, however, the recovery of the resolved alcohol involves the cleavage (hydrolytic or reductive) of the ester bond.

Other types of derivatives have been employed with success, e.g., hydrazone derivatives of carbonyls, molecular complexes, inclusion complexes (clathrates). There is great promise in the development of large chiral host molecules (chiral cages) specifically designed for the resolution of target racemates. The reader is warmly encouraged to consult (for further details and a full theoretical treatment) a clear and comprehensive review by Wilen on resolving agents and resolution published a few years ago [36].

Resolution by chromatography may fill preparative or analytical purposes. Globally, two distinct methods are used in chromatographic resolution, both of which are current and actively investigated. The direct chromatography is based on diastereoisomeric interactions between racemic solutes (now sometimes called selectands) and a chiral stationary phase (selector). Partial resolutions have been obtained in early liquid chromatography using solid chiral phases such as lactose, starch, quartz, or in paper chromatography (cellulose) (see, for example, sources in Refs. 44 and 45). Modern chiral stationary phases in liquid chromatography include Sephadex gels, cyclic ether host molecules attached to a support, optically active ion-exchange resins [45a], and chiral charge transfer agents absorbed on a silicic acid support [45]; the latter agents are also of promise in high-pressure liquid chromatography [46].

Optically active stationary phases have also proven to be of great utility in gas chromatography, either on capillary columns or on packed

TABLE 2 Common Resolving Agents

Acids	Bases
Tartaric acid	2-Amino-1-butanol
O,O'-Diacytartaric acids	Amphetamine
Tartranilic acids	Ephedrine
Malic acid	α-Phenylethylamine
Menthoxyacetic acid	α-(2-Naphthyl)ethylamine
Menthoxyacetyl chloride	α-Fenchylamine
Camphoric acid	Dehydroabietylamine
Camphor-10-sulfonic acid	Strychnine
3-Bromocamphor-9-sulfonic acid	Brucine
Pyroglutamic acid	Quinine
D-Arabonic acid	Quinidine
	Cinchonine
	Cinchonidine

columns (Table 3). Much work has been done on amino acid resolution by this technique [47].

The second method of chromatographic resolution involves derivatization with a chiral reagent followed by separation of the diastereoisomers on an achiral stationary phase. This method is especially well-suited for gas chromatography. Although physical separation by preparative gas chromatography is feasible, the method is mainly used for analytical purposes in biochemistry and drug metabolism studies where only microgram amounts of material are available. The reagents used are acylating agents forming stable amides with amines, or esters with alcohols. Some useful reagents are listed in Table 3; this matter has been discussed at length in a recent review [47].

Of particular interest appear diastereoisomeric amides and carbamates derived from chiral acids and alcohols, respectively [47b,47c]. In liquid chromatography, the diastereoisomers display a correlation between structure, stereochemistry, and elution order, thus allowing determination of absolute configuration.

TABLE 3 Some Compounds Used in Gas Chromatographic Resolution[a]

Chiral reagents

N-Trifluoroacetyl-(S)-prolyl chloride

(S)-(-)-N-Pentafluorobenzoylprolyl 1-imidazolidide

(+)-α-Methyl-α-methoxypentafluorophenylacetyl 1-imidazolidide

Drimanoyl chloride

(+)-Isoketopinyl chloride

Chiral stationary phases

Ureide of L-valine isopropyl ester

N-Trifluoroacetyl-L-valyl-L-valine cyclohexyl ester

N-Docosanoyl-L-valine tert-butylamide

Optically active lanthanoid complexes[b]

---

[a]From Testa and Jenner [47].

[b]From Golding et al. [47a].

The utility of gas and liquid chromatography also lies in the total resolution of enantiomers which can be achieved under good conditions. The method is, therefore, of value in the direct and reliable measurement of the optical purity of a given compound. Raban and Mislow have reviewed the theoretical bases of the method, together with many examples of applications [35].

12.6 Stereochemical Applications of Spectroscopic Methods and X-ray Crystallography

Spectroscopy and X-ray crystallography are methods of partial and total observation, respectively, of molecular structures. These methods are the major contributors to the sustained expansion of stereochemistry over the last two or three decades. Their importance and complexity are such that it would be mere dishonesty to summarize them in a few pages. Rather, the present section will briefly discuss some principles governing stereochemical applications of spectroscopic methods. By so doing, we hope to give a general understanding of what information the individual observation methods can provide, and of what information is inaccessible to them. Table 4 summarizes the key points of this discussion. Selected references of reviews and books are provided for the interested reader.

Ultraviolet (UV) spectroscopy is a typically achiral method, unable, therefore, to discriminate between enantiomeric molecules. As compared to IR and NMR spectra, for example, UV spectra contain relatively little structural information, and their use in conformational analysis is limited to some well-defined cases [48]. Deviation from planarity due to steric effects of substituents decreases the delocalization of π electrons in suitable molecules, and this is generally detected in UV spectra by a smaller λ_{max} and a smaller ϵ. UV spectroscopy is also of utility in discriminating between diastereoisomeric conformations such as the s-cis- and s-trans-forms in conjugated systems. Correlations with compounds of known geometry allows conformational assignment to be made in such cases [49].

The main stereochemical application of UV spectroscopy is the study of configurational diastereoisomers (discrimination and assignment of configuration). This is particularly true for π diastereoisomerism in conjugated systems. The effective length of the conjugated system will be greater for the trans- than for the cis-configuration. As a result, one often notes larger wavelengths and higher intensities of absorption for the trans- than for the cis-isomer [49].

Infrared (IR) spectroscopy is an achiral method of great value in the study of diastereoisomers. Because functional groups in diastereoisomeric molecules experience different molecular environments, they will also absorb at different frequencies. This effect arises mainly from differences in symmetry, and, to a lesser extent, from changes in nonbonded interactions [49]. While IR spectroscopy is potentially useful in discriminating any type of diastereoisomers and even in determining the composition of a mixture of diastereoisomers, it is of very limited help in assigning configurations. Indeed, it is usually extremely difficult to correlate with confidence shifts in absorption frequency with differences in molecular configurations. Assignments usually rest on comparison with analogous examples, a method not devoid of ambiguity.

IR spectroscopy has also contributed much in conformational analysis [48,49]. Absorption frequencies of groups differ in diastereoisomeric conformers. The conformation of alicyclic systems has been much-studied by this technique, since for a given group the equatorial vibration is almost always detected at a higher frequency than is the axial one.

Intramolecular H bonds can also be studied by IR spectroscopy due to the lower absorption frequency of hydrogen-bonded groups as compared to

TABLE 4 Significant Stereochemical Applications of Spectroscopic Methods

Method	Application
UV spectroscopy (conjugated systems)	Discrimination of π diastereoisomers and assignment of configuration Discrimination of diastereoisomeric conformers in some well-defined cases (s-trans- and s-cis-isomers, deviation from planarity)
IR spectroscopy	Discrimination of diastereoisomers Quantitative analysis of mixtures of diastereoisomers Conformational analysis of alicyclic molecules Assessment of intramolecular H bonds Differentiation between optical antipodes, racemates, and conglomerates
ORD and CD	Correlation of absolute configurations of chiral molecules Conformational studies of chiral molecules of known configuration
^{1}H-NMR spectroscopy	Discrimination of diastereoisomers and assignment of configuration (chemical shifts and coupling constants) Discrimination of diastereotopic protons Discrimination of diastereoisomeric conformers at low temperature, determination of the conformational free energy difference Determination of free energy of activation of intramolecular rate processes by dynamic NMR Calculation of populations of conformers from splitting patterns Discrimination of enantiomers by the use of chiral solvents and chiral LSR
^{13}C-NMR spectroscopy	Discrimination of diastereoisomers, assignment of configuration

TABLE 4 (continued)

Method	Application
^{13}C-NMR spectroscopy (continued)	Discrimination of conformers at low temperature, determination of the conformational free energy difference
	Determination of free energy of activation of intramolecular rate processes by dynamic NMR
	Determination of dihedral angles in rigid and flexibles molecules from C-H coupling constants
	Assessment of internal motions from ^{13}C relaxation times

unbonded ones. The detection and the exact magnitude of the frequency shift resulting from an intramolecular H bond can yield much conformational information. In conditions minimizing intermolecular bonds, the absence or presence of an intramolecular H bond can be proven, or a ratio of bonded to unbonded conformers approximated.

Section 12.5 has outlined some physical differences displayed in the solid state by the various mixtures of enantiomers. IR spectra of compounds in the solid state are sensitive to the crystalline form, and, as a consequence, will allow differentiation between optical antipodes, racemates, and conglomerates [50].

Unlike UV and IR spectroscopy, the chiroptical spectroscopic methods are typically chiral, as emphasized by this generic name. These methods include optical rotatory dispersion and circular dichroism, whose theoretical bases have been briefly presented in Section 12.3. The value of ORD and CD spectra (e.g., Figure 10) as opposed to single wavelength optical rotation lies in the measurement of Cotton effects. The latter are seen when $\epsilon_L - \epsilon_R = \Delta\epsilon \neq 0$ (see Section 12.3), and, therefore, imply two features in a molecule, namely, dissymmetry and a chromophore.

Cotton effects are exhibited by molecules having either an intrinsically dissymmetric chromophore (e.g., twisted biphenyls), or a dissymmetrically perturbated symmetric chromophore (e.g., chiral ketones). Of fundamental importance for stereochemical studies is the fact that the sign and the amplitude of the Cotton effect are related to the dissymmetry of the molecule. Configuration or conformation (either of them, but usually not both together) can be deduced from Cotton effect curves either by a

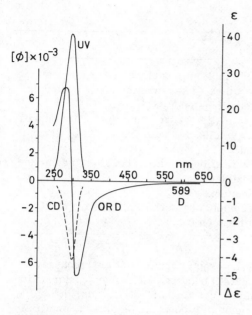

FIGURE 10 UV, ORD, CD spectra of a saturated ketone; the compound exhibits a simple negative Cotton effect for the n-π* transition as evidenced by the negative CD curve, and by the ORD curve (trough occurring at a higher wavelength than the peak). Note the levorotation at the sodium D-line.

theoretical treatment or by the popular <u>sector rules</u>. The latter are empirically or rationally deduced divisions of space around a chromophoric group; their occupation by atoms is considered to make a positive or a negative contribution to the observed Cotton effect.

The ketone chromphore has for long been the most studied one, and it was found that the so-called "octant rule" accounted for most results. The ketone octant rule is depicted in Figure 11, with a chair conformer of cyclohexanone in position. The attribution of atoms and groups into sectors corresponds to assigning approximate Cartesian coordinates; the <u>qualitative</u> contribution of each atom is obtained by multiplying the three "coordinates." For example, a substituent at C-3 (Figure 11) is located in the right-upper-rear octant (negative contribution), a 2-axial substituent brings a positive contribution, while a 2-equatorial substituent virtually does not contribute to the Cotton effect.

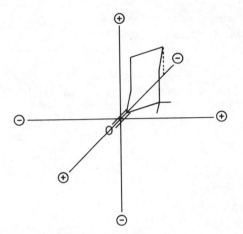

FIGURE 11 The octant rule for the ketone chromophore.

This qualitative interpretation is a mere approximation. It has been possible, in many cases, to determine the quantitative contribution of substituents to the observed Cotton effect ($\Delta\epsilon$). A few values are given in Table 5; these values are not absolute contributions, but differences between the contribution of the substituent and that of a hydrogen atom [50a]. Substituents whose contribution has the sign predicted by the octant rule are designated consignate; they are termed dissignate in the opposite case [50b]. Noteworthy is the dissignate behavior of fluorine, of some oxygenated groups, and of 3-axial substituents (Table 5). The contribution of ring junctions and fused rings to the n-π* Cotton effect of cyclohexanone has also been quantified [e.g., 50a].

Many other chromophores have been studied in connection with their Cotton effects, and several among them are presented in Table 6. The list is far from being exhaustive, but nevertheless gathers popular and less frequently studied chromophores. Useful derivatives are also given; they allow the absorption maximum to be shifted to higher wavelengths [e.g., 39,47, and sources therein; 48,51].

The ketone chromophore was the first studied because its absorption maximum (in the region of 300 nm) was readily accessible to the earlier spectropolarimeters and dichrographs. Modern instruments allow measurements down to 200 nm and below; this has resulted in easy observations of aromatic compounds. The benzene chromophore has proven to be of great interest in configurational and conformational studies. Based on

TABLE 5 The Contribution of Substituents of Cyclohexanone
(Chair Conformation) to the n-π* Cotton Effect[a]

| Substituent | $\delta \Delta \epsilon$ | | | |
	2-Equatorial	2-Axial	3-Equatorial	3-Axial
-CH$_3$	+0.2	+1.3	+0.55	+0.5; -0.1
-CH(CH$_3$)$_2$	+0.45[b]	+2.43[b]	+0.85	
-COOH	0.0		+0.47	-0.28
-N(CH$_3$)$_2$	+0.40	+0.92		
-OH	-0.8[c]	+0.96	+0.54	+0.23; -0.76
-OAc	0.0	+0.04	+0.20	-0.53
	0.65[c]	-0.7[c]		
-F		-1.2	-0.19	+0.03
-Cl		+3.5	+4.15	-0.41
-Br		+7.5; +4.5	+7.88	-0.45

[a]From Ripperger [50a]. Solvent: dioxane, unless otherwise indicated.
[b]Methanol
[c]Hexane

symmetry considerations developed by Schellman [52], Snatzke has proposed sector rules for the ^{1}L$_b$ (lowest energy transition, about 280 to 250 nm
and ^{1}L$_a$ (about 230 to 200 nm) bands of simple aromatic compounds were the
symmetric benzene chromophore is asymmetrically perturbated [52a, 52b].
The first step in the treatment is to consider chiral spheres in a molecule;
that chiral sphere which is nearest to the chromophore mainly determines
the Cotton effect (sign and magnitude). The first sphere is the aromatic
chromophore itself; it is chiral only in such compounds as hexahelicenes
(inherently dissymmetric chromophore, see Section 7.3) and leads to very
large $\Delta \epsilon$ values. The second chiral sphere is found in tetralins and

TABLE 6 Compounds Having a Chromophore of Interest in
Chiroptical Studies

Saturated carbonyl compounds

α, β - and β, γ -unsaturated ketones

Carboxylic acids

Esters and lactones

Amides and lactams

Olefins

Conjugated dienes

Monosubstituted benzene derivatives

Di- and polysubstituted benzene derivatives

Biphenyls

Helicenes

Derivatives of alcohols

    Esters
    Xanthates
    Thiourethanes
    Nitrous Esters

Derivatives of amines

    Dithiocarbamates
    N-Salicylidenes
    N-Phthaloyl derivatives
    Nitrosamines
    N-Chloramines

Derivatives of carboxylic acids

    Acylthioureas
    Thionamides

Derivatives of hydroxy and amino acids

    Copper complexes

Derivatives of olefins

    Osmate esters
    Episulfides

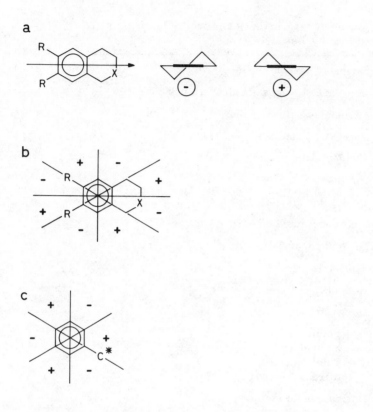

FIGURE 12 Sector rules for the ^1L_b-Cotton effect of the benzene chromophore (X = CR'$_2$ or NR'; R = H or OR"). From Snatzke and Ho [52a].
Reproduced with permission of the copyright holder.
(a) Rule for the second chiral sphere of tetralins (or tetrahydroisoquinolines); the conformation of the cyclohexene or tetrahydroazine ring
determines a negative or a positive Cotton effect as shown. (b) Rule for
the third chiral sphere of tetralins or tetrahydroisoquinolines (upper
sectors shown, the lower sectors have opposite signs). Substituents
attached to the second chiral sphere (= attached to the alicycle) have a
positive or a negative contribution to the Cotton effect depending on their
location. (c) Rule for the third chiral sphere of monosubstituted benzene
derivatives (upper sectors shown, the lower sectors have opposite signs).
Groups attached to the C* have positive or negative contributions depending
on their location.

tetrahydroisoquinolines (see Figure 12a), while the third chiral sphere is represented by the substituents on the alicyclic moiety of, for example, tetralins (Figure 12b) or on the chiral center of asymmetric monosubstituted benzene derivatives (Figure 12c). The more the chiral sphere is remote from the chromophore, the less it contributes to the Cotton effect. Figure 12 gives the signs of the contributions of second and third chiral sphere substituents to the ^1L_b band Cotton effect [52a]; the signs associated with the ^1L_a band are not considered definitive and have not been given here. Other sector rules apply to polysubstituted benzene derivatives [52b, 52c].

For chiral compounds of rigid structure or of well-known conformation, the chiroptical methods allow assignment of absolute configuration by correlations with closely related analogs, by the use of sector rules (a more general and less reliable means of correlation), and sometimes also by a direct, noncorrelative, theoretical treatment. Further, the sector rules are of great utility in conformational studies of chiral compounds of known absolute configuration, especially in following solvent- and temperature-induced conformational changes.

The spectroscopic method which has contributed the most to the recent development of stereochemistry is certainly nuclear magnetic resonance (NMR) spectroscopy. The following pages will present the stereochemical applications of ^{1}H-NMR and ^{13}C-NMR spectroscopy. The basic knowledge of NMR can be found in numerous books [e.g., 53-55] and will not be discussed here. Several outstanding general reviews on the stereochemical applications ^{1}H-NMR have been published [e.g., 48, 56, 57].

The interpretation of ^{1}H-NMR spectra rests on two main approaches; one makes use of chemical shifts, the other of coupling constants. Due to differences in spatial relationships of atoms and to the magnetic anisotropic shielding effects of many functional groups, protons in diastereoisomeric environments will experience nonequivalent electromagnetic influences resulting in differences in chemical shifts. The discrimination of diastereoisomers is well-known and was among the first stereochemical applications of the method. This is well-illustrated by the diastereoisomers of stilbene (XXXII); indeed, the two olefinic protons are more

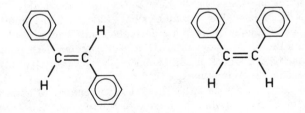

XXXIIa   XXXIIb

deshielded in the trans-isomer (XXXIIa) than in the cis-isomer (XXXIIb), $\delta = 6.99$ and 6.49 ppm, respectively [56]. Another example is provided by 4-tert-butylbromocyclohexane; the proton on the carbon atom carrying the bromine substituent is axial in the trans-diastereoisomer and resonates at $\delta = 3.30$ ppm, whereas in the cis-isomer it is equatorial and resonates at $\delta = 2.67$ ppm [48]. These variations of the chemical shifts are usually rationalized in terms of shielding effects and allow reliable assignments of configuration to be made.

Besides distinguishing protons in diastereoisomeric molecules as in the above examples, ^{1}H-NMR spectroscopy allows discrimination of diastereotopic protons. Bromoethylene (XXXIII) shows two such protons; H' and H'' resonate at $\delta = 5.75$ and 5.83 ppm, respectively [57]. Chemical shift differences typically in the range 0.02 to 0.10 ppm are also found between diastereotopic protons on a prochiral center, e.g., of the type W-CH_2-CXYZ. An illuminating review has been published recently on the chemical shift nonequivalence in prochiral groups [58].

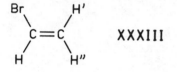 **XXXIII**

The discrimination of diastereoisomeric conformers by the chemical shift method is usually not feasible at room temperature because the rate of interconversion is fast compared to the NMR time scale. As a consequence, one does not observe two distinct signals when two conformers are present, but rather a single peak which is an average for the two isomers. At a sufficiently low temperature, the exchange will be slow compared to the NMR time scale, and the spectrum will display an individual resonance peak for each conformer. The ratio of the peak areas gives the conformational ratio. The free energy difference between the two conformers can then be calculated (see Section 9.1).

If, at higher temperatures, two diastereoisomeric conformers give a single average resonance peak for a given group, and, if at sufficiently low temperatures the signals are separated into distinct peaks, then the stepwise temperature increase should show the two peaks broaden, overlap, coalesce, and become a sharp, single peak. Indeed, this transition is observed to occur gradually over a relatively small temperature interval and constitutes the basis of dynamic NMR spectroscopy. Of particular significance is the temperature of coalescence (T_c). From an analysis of the spectra at T_c, the rate constant for interconversion (k_c) can be determined; the latter, in turn, allows calculation of the free energy of activation (energy barrier of the process) by means of Eyring equation (Section 4.1,

Eq. 1). The determination of the rate constant k_c is the critical step in dynamic NMR; treatment of various levels of sophistication can be chosen, but it appears that the simple and classical treatments are often misleading. Therefore, this point will not be discussed further, and the reader is referred to in-depth treatments [59,60]. The stereochemical applications of dynamic NMR spectroscopy include the study of hindered rotations, inversion of lone electron pairs (inversion processes at atomic centers), and ring inversions [59]. The temperature range accessible to current spectrometers spans from -150 to +200°C, which corresponds to free energy barriers ranking from 23 to 110 kJ/mol (5.5 to 26 kcal/mol).

Besides chemical shifts, another source of stereochemical information in NMR spectra are the coupling constants, symbolized by J; these are a measure of the spin-spin interactions between atoms A and B, resulting in splitting patterns of the resonance signals. The coupling constants $J_{HH'}$ between vicinal protons (H-C-C-H') will be the only one discussed here. Four molecular parameters influence the magnitude of $J_{HH'}$, namely, (see, for example, Refs. 61 and 62):

1. The dihedral angle θ between the protons; this dependence is of fundamental importance in stereochemistry. The relationship has been shown by Karplus to be of the general form

$$J_{HH'} = a \cdot \cos^2 \theta - c \qquad (10)$$

   where the value of a is dependent on the system and is larger for $0° \leqslant \theta \leqslant 90°$ than it is for $90° \leqslant \theta \leqslant 180°$; c is a constant which is generally very small or nil.

2. The electronegativity of substituents; it is generally found that an increased electronegativity of the substituents decreases $J_{HH'}$, i.e., a in Eq. (10) will have smaller values.

3. The C-C-H bond angles; $J_{HH'}$ should decrease as α (C-C-H) and α' (C-C-H') increase.

4. The bond lengths; although this dependence is of little practical significance, $J_{HH'}$ is expected to decrease with increasing C-C bond length.

The determination of $J_{HH'}$ is of obvious value in ascertaining the configuration of stereogenic elements such as alicyclic ring junctions and disubstitution of alicycles (e.g., cis-trans and endo-exo diastereoisomerism). Even a rough determination of θ from $J_{HH'}$ will allow one to make such configurational assignments.

Even more important are the conformational applications of $J_{HH'}$. For a flexible molecule existing as two or several conformers (whose interconversion is fast compared to the NMR time scale), the observed $J_{HH'}$ will be the resultant of the weighted contributions of all conformers present. Theoretical treatments have been conceived allowing the calculation of populations of conformers from the splitting patterns [63-65; see also Ref. 57 for numerous examples].

The range of stereochemical applications of ^{1}H-NMR spectroscopy has been greatly increased by the technique of diastereoisomeric interactions. Optically active solvents such as 2,2,2-trifluoro-1-phenylethanol give rise to small differences in the chemical shifts (smaller than 0.1 ppm) of some resonance peaks of enantiomeric solutes [57]. But the chiral handles which have added so much range to NMR spectroscopy are the chiral lanthanide shift reagents. A lanthanide shift reagent (LSR) consists of a six-coordinate metal complex which, by expanding its coordination sphere, accepts further ligands. Heteroatoms exhibiting some degree of Lewis basicity provide the required ligands. Paramagnetic lanthanide complexes are formed, and, as a result, large pseudocontact shifts (lanthanide-induced shifts, LIS, designated $\Delta\delta$) are induced in nuclei close to the heteroatoms [66,67].

Achiral LSR have found many applications in stereochemical studies, such as increased resolution and assignment of signals from diastereoisomeric and diastereotopic protons. Also, they are of utility and great promise in conformational studies [68-70]. Chiral LSR, on the other hand, raise considerable interest in the discrimination of enantiomers. Useful reagents include tris [3-(tert-butylhydroxymethylene)-d-camphorato] europium(III) [Eu(t-bhmc)₃, XXXIV], tris (3-trifluoroacetyl-d-camphorato)europium(III) [Eu(facam)₃, XXXV] and its praseodymium analog Pr(facam)₃, and tris(3-heptafluorobutyryl-d-camphorato)europium-(III) [Eu(hfbc)₃, XXXVI] and Pr(hfbc)₃. In the presence of such reagents, corresponding protons in enantiomeric molecules experience differences in the pseudocontact shifts (designated $\Delta\Delta\delta$) which are usually in the range 0.00 to 0.20 ppm, but not infrequently much larger [47,66,71-73].

Given a proper reagent, optimal ranges in the molar ratio of LSR: substrate can thus be found which allow maximal separation of enantiomeric

$R = C(CH_3)_3$ **XXXIV**

$R = CF_3$ **XXXV**

$R = nC_3F_7$ **XXXVI**

signals and quantitative discrimination of two enantiomers. Amusingly, some people have coined this technique "polarimetry by NMR."

A drawback of chiral LSR is that the relative shifting of the (R)- and (S)-enantiomers is currently difficult to predict, and thus not well-suited in many cases for the determination of absolute configurations.

Related to the chiral solvents and LSR are the chiral solvating agents recently investigated. These compounds are designed for selectivity toward some classes of substrates, with which they build diastereoisomeric complexes by two- or three-point interactions. For example, optically active 2,2,2-trifluoro-1-(9-anthryl)ethanol interacts with chiral lactones, allowing determination of their enantiomeric purity and absolute configuration by ^{1}H- and ^{13}C-NMR spectroscopy [73a].

Carbon-13 NMR spectroscopy is catching up fast with ^{1}H-NMR as regards stereochemical applications [74]. A masterly review has been published recently on stereochemical aspects of ^{13}C-NMR spectroscopy [75]; a few salient points will now be concisely discussed.

Two major parameters of ^{13}C-NMR spectra are ^{13}C shieldings and [13]C-^{1}H coupling constants, both of which are a mine of stereochemical information. Carbon-13 spectra recorded with complete proton decoupling usually show separate resolved signals for each individual carbon in not too complex molecules. The different environments of corresponding carbons in diastereoisomeric molecules will result in chemical shift differences ($\Delta\delta$); these are considerably larger than the analogous differences found in ^{1}H-NMR spectra. For example, a methyl group will show different δ_C in endo- or exo-, axial- or equatorial-, cis- or trans-positions; in fact, all or almost all carbon atoms will exhibit $\Delta\delta$ values of varying magnitude in diastereoisomeric molecules. The discrimination of diastereoisomers is thus very good as a rule, and allows configurational assignments based on direct comparison with related compounds, or indirect comparison by means of rules or models. An extension of this method involves the derivatization of chiral compounds with an optically active reagent; recently, chiral ketones have been transformed into their diastereoisomeric cyclic ketals, and their enantiomeric purity determined using ^{13}C-NMR spectroscopy [75a].

Diastereosiomeric conformers are also discriminated, again quite often with a much increased resolution power as compared to ^{1}H-NMR spectroscopy. Accurate conformational ratios have thus been measured by low-temperature ^{13}C-NMR spectroscopy [e.g., 76].

Dynamic ^{13}C-NMR spectroscopy is based on the same principles and spectral behavior as ^{1}H-NMR spectroscopy. Here also, free energies of activation are determined from coalescence temperatures. The two techniques usually yield convergent results; the contrary means that two different processes have been looked at!

The coupling constant between adjacent carbon and hydrogen atoms (^{13}C-^{1}H coupling constants) are known to increase with the electronegativity of substituents or heteroatoms, especially in the α position; they also increase with increasing s character of the hybrid bonding orbital of the carbon atom. The latter parameter is of a geometrical rather than stereochemical nature, since it is related to the bond angles at the carbon atom.

Much less studied, but of great stereochemical promise are the long range C-H coupling constants, especially the vicinal coupling constants (J_{CCCH}). A dihedral angle dependence for J_{CCCH} is now well documented [75].

Relaxation times in ^{13}C-NMR spectroscopy also provide structural information, as well as insight into dynamic intramolecular processes. For example, the internal motions of individual carbons can be compared, allowing the quantitative assessment of flexibility and rigidity within a molecule [75]. The rapidly expanding topic of relaxation times will not be discussed further here.

X-ray crystal-structure analysis is a potent technique for the direct observation of molecules. The topic of crystallography and stereochemistry has been well reviewed [e.g., 77-81]. The discussion to follow will briefly mention the stereochemical applications of the technique, while no theoretical background will be given.

Ordinary X-ray analysis gives detailed information on the relative spatial disposition of atoms in a molecule. If the molecule contains two or more elements of chirality, their relative configuration will be established. The method will thus discriminate between diastereoisomers. The absolute configuration of chiral elements, however, cannot be assessed by ordinary X-ray analysis, which means that enantiomeric molecules cannot be discriminated. If the absolute configuration of one dissymmetric group in the molecule is known, the absolute configurations of the other centers will then be determined relatively to the first.

Anomalous X-ray scattering (anomalous dispersion, differential Bijvoet X-ray analysis) is the ultimate technique for the determination of absolute configuration. Indeed, anomalous dispersion, as opposed to ordinary X-ray analysis, will assign the absolute configuration of an optically active molecule. The epoch-making work of Bijvoet and co-workers in 1951 [82] provided the first unambiguous demonstration of the correctness of the Fischer convention (see Section 5.3) by determining the absolute configuration of (2R;3R)-(+)-tartaric acid (as the Na, Rb salt). The absolute configurations of a great number of chiral molecules have now been established by anomalous X-ray crystallography. The configuration of almost all other chiral molecules has been determined by stereochemical correlations ultimately based upon X-ray crystallography [39]. The exceptions mentioned refer to allotments of absolute configuration by circular dichroism [39].

Because X-ray crystallography is a method of total structure determination, it will allow determination not only of configuration, but also of total molecular geometry (bond lengths, bond angles, dihedral angles). X-ray dispersion is certainly the most accurate method for the determination of conformation. However, the major limitation lies in the fact that the structures are determined for compounds in the crystalline state [48]. The conformation of molecules in the crystal may be quite different from that in solution, and it is the latter which bears the main relevance to chemical, biochemical and pharmacological problems. Although provocative, it is not totally absurd to consider molecules in crystals as "fossilized molecules" [78].

Other methods yielding stereochemical information include [48,81] neutron diffraction, electron diffraction, microwave spectroscopy [83,84], ultrasonic absorption, and vibrational spectroscopy [85]. The stereochemical aspects of mass spectrometry arise increasing interest and have been thoroughly reviewed [86]; they will not be discussed here.

12.7 Theoretical Conformational Analysis

Any discussion of stereochemical methodology would not be complete if it were restricted solely to experimental techniques. Indeed, theoretical analysis of organic molecules are extremely useful in studying energy variations associated with changes in such structural parameters as dihedral angles, bond lengths, and bond angles. Among these, the dihedral angles are major stereochemical parameters; the following discussion will be restricted to their analysis in terms of energy variations (theoretical conformational analysis).

The general principle of conformational calculations consists of calculating the potential energy of each individual conformer. The various theoretical methods used to calculate the potential energy will be discussed later, but let us consider first the presentation of results. As a first example, we consider a compound with a single degree of conformational freedom, namely, the dihedral angle θ. This angle is rotated from $0°$ to $360°$ by increments of $360°/n$, and the potential energy of the molecule is calculated for each conformer thus defined. If n is small, only a few conformers will be considered (e.g., the eclipsed, gauche-, and trans-forms), and the results will allow a comparison of their relative stability. If more conformers are studied (n increased) a plot of potential energy versus θ

can be drawn, allowing a better visualization and understanding of the conformational behavior of the molecule studied. Some examples of such plots can be found in Chapters 9 and 10.

Plots of potential energy versus dihedral angle θ can be recalculated to yield <u>plots of relative population of conformers versus θ</u>, making the display even more eloquent to the eye. Figure 13 compares two such plots. The method by which Figure 13a was obtained is irrelevant here, the purpose of Figure 13 being only to show to the reader how usefully the two plots 13a and 13b complement each other, and how well they visually describe the conformational phenomenon. In view of this, it is unfortunate

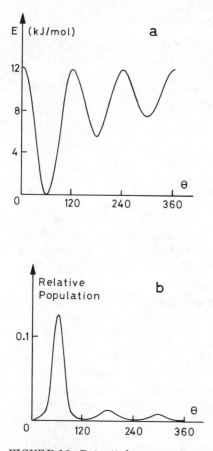

FIGURE 13 Potential energy diagram (a) and population profile (b) of rotation about the C_α-C_β bond of noradrenaline. From Roberts [87].

that plots of relative population versus θ are so rarely found in the literature.

For a molecule with two degrees of conformational freedom, one of the dihedral angles (e.g., θ_2) can be assigned a given value, while θ_1 is rotated as previously discussed. A plot of potential energy versus θ can then be drawn, which is, however, valid for one value or θ_2 only. The comparison of several such plots (for selected values of θ_2) is not very informative. In recent years, conformational energy maps have become more and more frequent in the chemical and especially medicinal chemical literature. Such maps are obtained by simultaneously varying the two dihedral angles θ_1 and θ_2, and by plotting the results as isoenergy contours; each contour represents a potential energy increment of a fixed value. The simultaneous variation of θ_1 and θ_2 implies the calculation of many points (for 10° increments, i.e., a 10° grid, there will be 36^2 points, also called intersections). An example of a conformational energy map is shown in Figure 14; the similarity with a topographical map and its altitude levels is evident.

Conformational energy maps can be transformed to yield percentage maps, in which a single contour indicates the regions of conformation within which any given percentage of molecules can be expected to be found. Such maps highlight differences in conformational preference between analogous compounds [88]. The percentage map calculated from Figure 14 is displayed in Figure 15.

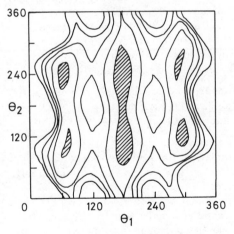

FIGURE 14 Conformational energy map for histamine monocation; the contours represent 0.1 eV increments and the low-energy fields are hatched (θ_1 and θ_2 are the dihedral angles of the C_α-C_β and C_β-C_γ bonds, respectively). Adapted from Richards and Ganellin [88].

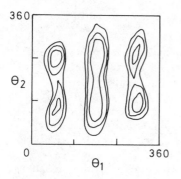

FIGURE 15 Percentage map for the histamine monocation, as calculated from the conformational map of Figure 14; outer, median and inner contours represent probabilities of 99, 95, and 75%, respectively. Adapted from Richards and Ganellin [88].

Conformational energy maps such as the one in Figure 14 are three-dimensional representations reduced to two dimensions. It is also possible to find in the literature perspective drawings of the three-dimensional diagrams.

For compounds with more than two degrees of conformational freedom, the simultaneous variation of all dihedral angles would be prohibitive in terms of computer time, and the results could not be displayed graphically. In such cases, only the two most important (in the context of the study) dihedral angles are varied, and the others are fixed in conformations assumed to be preferred in the light of preliminary calculations. Alternatively, some portions of space or hyperspace around a minimum in a conformational map can be mapped by varying the other dihedral angles.

Let us now briefly consider the theoretical methods used to calculate the conformational behavior of molecules. These methods are listed in Table 7; they have been discussed in an interesting review article [89].

The classical methods (molecular mechanical methods) calculate conformational energies by summing atom-atom interactions (bonded and nonbonded terms). Empirical potential energy function (involving empirically determined parameters) are used to describe the conformational behavior of a molecule [84]. Of particular importance in such calculations are the repulsive terms of electron clouds around nonbonded atoms. Other nonbonded terms include ionic interactions and hydrogen bonding. Elaborate calculations will also consider bond length and bond angle variations. All or several terms of a given coherent set are thus added, and the differences in energies of conformers are obtained by adequate substractions.

TABLE 7 Theoretical Methods of Conformational Analysis

Classical (empirical) methods

Quantum mechanical methods

Semiempirical methods

π Electrons only

HMO: Hückel Molecular Orbital Theory
PPP: Pariser-Parr-Pople Theory

σ and π Electrons

(All-electron and all-valence-electron theories)

EHT: Extended Hückel MO Theory
CNDO: Complete Neglect of Differential Overlap
INDO: Intermediate NDO
MINDO: Modified INDO
PCILO: Perturbative Configuration Interactions using
           Localized Orbitals

Nonempirical methods (ab initio MO methods)

Minimal basis: STO-3G: Slater-type orbital at the
                                       3-Gaussian level
Split-valence basis: 4-31G
Polarized basis: 6-31G*

The classical methods are cheap in terms of computer time, they can be readily refined according to needs, and they are easily applicable to large molecules. On the other hand, there are difficulties in case of heteroatoms, and the many oversimplifications call for a lucid caution in the interpretation of results. But fair or good agreements with other methods are common, and we believe the empirical methods to provide a valuable and sometimes underrated tool in conformational analysis [89,89a].

The quantum mechanical methods have become extremely popular in conformational analysis, but to a variable extent. The π-electron theories (HMO and PPP, Table 7) are obviously applicable to π systems only, since they neglect all but one (p-π) orbital on each atom [90]; their utility in conformational analysis is quite limited [89] and does not call for further discussion in this limited space.

Most of the current theoretical work in conformational studies involves the all-valence and all-valence-electron semiempirical methods, and the nonempirical theories [89-91]. It is generally recognized that their accuracy in predicting conformational behaviors increases from top to bottom in the series:

EHT
CNDO
INDO, MINDO
PCILO
ab initio, minimal basis
ab initio, extended bases

The EHT is a simple and relatively cheap method. It appears to correctly predict the preferred conformations of molecules in which steric repulsion and delocalization are predominant [92]. The calculated barriers of rotation, on the other hand, are unrealistic, one of the reasons being that the steric interaction at long distances between nonbonded-atoms are badly overestimated. Further, the EHT method fails in accounting for electrostatic properties. As a result, the conformational behavior of molecules with marked charge separation, and of charged molecules, is very poorly reproduced. Fortuitious agreements with experimental studies in solution arise from the fact that solvation shells decrease electrostatic effects in polar solvents [92]. Globally, the EHT method thus appears as a little reliable conformational tool [89,91,92].

The NDO methods are less empirical than the Hückel models, but the neglect of the overlap term means that the short-distance repulsive effects are not, or not sufficiently, taken into account. The missing contributions are mimicked by delocalization terms [92]. Globally, however, the conformational results obtained with the NDO methods lack general reliability. Rotation barriers, in particular, are often unrealistic [93].

The PCILO method has been tailor-made for conformational studies, for which it appears well-suited. Indeed, the method is an undisputed improvement over the NDO approximations. The conformational results are considered reliable in terms of both preferred conformations and rotation barriers, since they are usually in good agreement with experimental data and with theoretical results obtained by ab initio calculations.

Another advantage of the PCILO over the NDO methods is that the results are obtained with far less computer time. The reliability and speed of PCILO make it quite useful for the calculation of explicit and meaningful conformational energy maps [93].

The ab initio methods consider all electrons, including the core electrons, and no integral is ignored or replaced by empirical parameters. In
these theories, each individual orbital is approximated as a linear combination of a set of given functions (basis functions). When there are just
enough functions to permit a description of the ground states of the separate
atoms, the basis set is said to be minimal. More sophisticated methods
extend the basis (extended basis sets) by introducing additional basis functions [94,95].

In terms of conformational studies, the accuracy of the ab initio methods increases with extending basis. But the computer time needed also
increases and quickly becomes prohibitive for anything but small molecules.
Therefore, the extension of the basis set until close to the Hartree-Fock
limit (see Ref. 90) is restricted to molecules with only a few atoms. But
the Hartree-Fock theory is itself an approximation, and therefore not fully
adequate to describe reality. Theoreticians have extended the range of
models beyond the Hartree-Fock limit, the goal being to approach the exact
solution of the nonrelativistic Schrödinger equation [90,95].

In practice, however, even the minimal basis set ab initio method
(STO-3G) yields results which approximate reality with good reliability.
Only in the case of highly accurate works or correlations is this method
judged as insufficient.

A drawback of the nonempirical methods is the computer time needed
for any calculation, as already mentioned. This means that these methods
cannot be used to calculate full conformational energy maps of even
relatively small molecules. Instead, they are used to check some interesting points (e.g., minima and barriers) in a conformational map obtained by
cheaper methods (e.g., PCILO, Ref. 93).

Up to this point, the discussion was centered around the theoretical
methods, the molecules being studied, and the results. No mention, however, was made to the environment of the investigated molecules. That the
environment is a critical factor in theoretical conformational studies has
become more and more evident in recent years. Theoretical studies
usually consider the molecules as isolated entities, i.e., in a vacuum.
Experimental studies, on the other hand, show that the conformational
behavior of a molecule is sensitive to environments such as the crystalline
state or solvents of different polarities. In view of these influences, the
positive agreements often found between theoretical and experimental
studies are unexpected or even puzzling, if they are not fortuitous.

To put theoretical conformational analysis on a firmer basis, solvation
is now being taken into account, and most frequently hydration. Several

models have already been used (e.g., hydration shells), but a simple and elegant method consists of fixing water molecules in the most favorable hydration sites and calculating the conformational maps of the new "supermolecules." The most favorable hydration sites are determined by careful preliminary studies using, e.g., an ab initio method [93,96]. No doubt further developments of solvation models will contribute significantly to the bright future of theoretical conformational analysis.

References

1. A. S. Garay, L. Keszthelyi, I. Demeter, and P. Hrasko, Nature, 250:332 (1974).

2. J. L. Abernethy, J. Chem. Educ., 49:455 (1972).

3. K. Mislow, Introduction to Stereochemistry, Benjamin, New York, 1966, pp. 52-54, 119-120.

4. A. Horeau and J. P. Guetté, Tetrahedron, 30:1923 (1974).

5. M. J. P. Harger, J. Chem. Soc. Chem. Comm., 555 (1976).

6. K. Mislow, Introduction to Stereochemistry, Benjamin, New York, 1966, pp. 54-61.

6a. H. Wynberg and B. Feringa, Tetrahedron, 32:2831 (1976).

7. E. L. Eliel, Stereochemistry of Carbon Compounds, McGraw-Hill, New York, 1962, pp. 7-9, 398-401.

8. E. L. Eliel, Stereochemistry of Carbon Compounds, McGraw-Hill, New York, 1962, pp. 398-412.

9. H. Wynberg, G. L. Hekkert, J. P. M. Houbiers, and H. W. Bosch, J. Am. Chem. Soc., 87:2635 (1965).

9a. H. Wynberg and L. A. Hulshof, Tetrahedron, 30:1775 (1974).

10. J. H. Brewster, Helix models of optical activity, in Topics in Stereochemistry, vol. 2 (edited by N. L. Allinger and E. L. Eliel), Wiley, New York, 1967, pp. 1-72.

11. J. L. Carlos, Jr., J. Chem. Educ., 45:248 (1968).

12. M. M. Harris, The study of optically labile compounds, in Progress in Stereochemistry, vol. 2 (edited by W. Klyne and P. B. D. de la Mare), Butterworths, London, 1958, pp. 157-195.

13. J. A. Mills and W. Klyne, The correlation of configurations, in Progress in Stereochemistry, vol. 1 (edited by W. Klyne), Butterworths, London, 1954, pp. 177-222.

14. K. Freudenberg, Ber. Deutsch. Chem. Ges., 66:177 (1933).

15. E. L. Eliel, Stereochemistry of Carbon Compounds, McGraw-Hill, New York, 1962, pp. 434-437.

16. P. Jenner and B. Testa, Drug Metab. Rev., 2:117 (1973).

17. E. L. Eliel, Stereochemistry of Carbon Compounds, McGraw-Hill, New York, 1962, pp. 151-152, 237-239.

17a. N. S. Zefirov, Tetrahedron, 33:2719 (1977).

18. P. Sykes, A Guidebook to Mechanism in Organic Chemistry, 3d ed., Longman, London, 1970, pp. 156-177.

19. E. L. Eliel, Stereochemistry of Carbon Compounds, McGraw-Hill, New York, 1962, pp. 295, 346-368.

20. P. Sykes, A Guidebook to Mechanism in Organic Chemistry, 3d ed., Longman, London, 1970, pp. 214-218.

21. P. Sykes, A Guidebook to Mechanism in Organic Chemistry, 3d ed., Longman, London, 1970, pp. 81-88.

22. P. B. D. de la Mare, Stereochemical factors in reaction mechanism and kinetics, in Progress in Stereochemistry, vol. 1 (edited by W. Klyne), Butterworths, London, 1954, pp. 90-125.

23. J. D. Morrison and H. S. Mosher, Asymmetric Organic Reactions, Prentice-Hall, Englewood Cliffs, New Jersey, 1971.

23a. L. Salem, J. Am. Chem. Soc., 95:94 (1973).

24. R. Altschul, P. Bernstein, and S. G. Cohen, J. Am. Chem. Soc., 78:5091 (1956).

25. J. S. Birtwistle, K. Lee, J. D. Morrison, W. A. Sanderson, and H. S. Mosher, J. Org. Chem., 29:37 (1964).

25a. F. Weygand, W. Steglich, and X. Barocio de la Lama, Tetrahedron, Suppl. 8, part I, p. 9 (1966).

26. D. J. Cram and F. A. Abd Elhafez, J. Am. Chem. Soc., 74:5828 (1952).

26a. D. J. Cram and D. R. Wilson, J. Am. Chem. Soc., 85:1245 (1963).

26b. G. J. Karabatsos, J. Am. Chem. Soc., 89:1367 (1967).

26c. M. Chérest, H. Felkin, and N. Prudent, Tetrahedron Lett., 2199 (1968).

26d. M. Chérest and H. Felkin, Tetrahedron Lett., 2205 (1968).

26e. V. Prelog, Helv. Chim. Acta, 36:308 (1953).

26f.   V. Prelog, Bull. Soc. Chim. France, 987 (1956).

26g.   E. Ruch and I. Ugi, The stereochemical analogy model: a mathe-
        matical theory of dynamic stereochemistry, in Topics in Stereo-
        chemistry, vol. 4 (edited by E. L. Eliel and N. L. Allinger), Wiley,
        New York, 1969, pp. 99-125.

27.   H. R. Levy, P. Talalay and B. Vennesland, The steric course of
        enzymic reactions at meso carbon atoms: application of H-isotopes,
        in Progress in Stereochemistry, vol. 3 (edited by P. B. D. de la
        Mare and W. Klyne), Butterworths, London, 1962, pp. 299-349.

28.   D. Arigoni and E. L. Eliel, Chirality due to the presence of hydrogen
        isotopes at noncyclic positions, in Topics in Stereochemistry, vol. 4
        (edited by E. L. Eliel and N. L. Allinger), Wiley, New York, 1969,
        pp. 127-243.

29.   W. L. Alworth, Stereochemistry and Its Application in Biochemistry,
        Wiley, New York, 1972.

30.   J. W. Cornforth, Tetrahedron, 30:1515 (1974).

31.   B. Vennesland, Stereospecificity in biology, in Topics in Current
        Chemistry, vol. 48, Springer, Berlin, 1974, pp. 39-65.

32.   K. R. Hanson and I. A. Rose, Acc. Chem. Res., 8:1 (1975).

33.   K. R. Hanson, J. Biol. Chem., 250:8309 (1975).

33a.   K. R. Hanson, Annu. Rev. Biochem., 45:307 (1976).

34.   T. H. Elliott, E. Jacob, and R. C. C. Tao, J. Pharm. Pharmacol.,
        21:561 (1969).

34a.   A. Schoofs and A. Horeau, Tetrahedron Lett., 3259 (1977).

35.   M. Raban and K. Mislow, Modern methods for the determination of
        optical purity, in Topics in Stereochemistry, vol. 2 (edited by
        N. L. Allinger and E. L. Eliel), Wiley, New York, 1967,
        pp. 199-230.

36.   S. Wilen, Resolving agents and resolutions in organic chemistry, in
        Topics in Stereochemistry, vol. 6 (edited by N. L. Allinger and
        E. L. Eliel), Wiley, New York, 1971, pp. 107-176.

37.   W. H. Hartung and J. Andrako, J. Pharm. Sci., 50:805 (1961).

38.   W. Klyne and J. Buckingham, Atlas of Stereochemistry: Absolute
        Configurations of Organic Molecules, Chapman and Hall, London,
        1974.

39.   W. Klyne and P. M. Scopes, Stereochemical correlations, in

Progress in Stereochemistry, vol. 4 (edited by B. J. Aylett and M. M. Harris), Butterworths, London, 1969, pp. 97-166.

40. W. Klyne and P. M. Scopes, Farmaco Ed. Sci., 24:533 (1969).

41. G. Krow, The determination of absolute configuration of planar and axially dissymmetric molecules, in Topics in Stereochemistry, vol. 5 (edited by E. L. Eliel and N. L. Allinger), Wiley, New York, 1970, pp. 31-68.

42. E. L. Eliel, Stereochemistry of Carbon Compounds, McGraw-Hill, New York, 1962, pp. 43-47.

43. A. Fredga, Bull. Soc. Chim. France, 173 (1973).

43a. H. Wynberg, Acc. Chem. Res., 4:65 (1971).

44. E. L. Eliel, Stereochemistry of Carbon Compounds, McGraw-Hill, New York, 1962, pp. 61-62.

45. W. H. Pirkle and D. L. Sikkenga, J. Chromatog., 123:400 (1976).

45a. B. J. Mehta and N. Krishnaswamy, J. Chromatog., 135:455 (1977).

46. F. Mikeš, G. Boshart, and E. Gil-Av, J. Chromatog., 122:205 (1976).

47. B. Testa and P. Jenner, Stereochemical methodology, in Drug Fate and Metabolism: Methods and Techniques, vol. 2 (edited by E. R. Garrett and J. Hirtz), Dekker, New York, 1978, pp. 143-193.

47a. B. T. Golding, P. J. Sellars, and A. Kee Wong, J. Chem. Soc. Chem. Comm., 570 (1977).

47b. W. H. Pirkle and J. R. Hauske, J. Org. Chem., 42:1839 (1977).

47c. G. Helmchen, H. Völter, and W. Schühle, Tetrahedron Lett., 1417 (1977).

48. E. L. Eliel, N. L. Allinger, S. J. Angyal, and G. A. Morrison, Conformational Analysis, Wiley, New York, 1965, chap. 3.

49. E. A. Braude and E. S. Waight, The relationship between the stereochemistry and spectroscopic properties of organic compounds, in Progress in Stereochemistry, vol. 1 (edited by W. Klyne), Butterworths, London, 1954, pp. 126-176.

50. G. Wirzing, Z. Anal. Chem., 267:1 (1973).

50a. H. Ripperger, Z. Chem., 17:250 (1977).

50b. W. Klyne and D. N. Kirk, Tetrahedron Lett., 1483 (1973).

51. K. Mislow, Introduction to Stereochemistry, Benjamin, New York, 1966, pp. 62-67, 155-161.

52. J. A. Schellman, Chem. Rev., 75:323 (1975).

52a. G. Snatzke and P. C. Ho, Tetrahedron, 27:3645 (1971).

52b. G. Snatzke, M. Kajtár, and F. Werner-Zamojska, Tetrahedron, 28:281 (1972).

52c. J. Sagiv, Tetrahedron, 33:2315 (1977).

53. N. S. Bhacca and D. H. Williams, Applications of NMR Spectroscopy in Organic Chemistry, Holden-Day, San Francisco, 1964.

54. L. M. Jackman and S. Sternhell, Applications of Nuclear Magnetic Resonance Spectroscopy in Organic Chemistry, 2d ed., Pergamon, Oxford, 1969.

55. J. W. Emsley, J. Feeney, and L. H. Sutcliffe, High Resolution Nuclear Magnetic Resonance Spectroscopy, vols. 1 and 2, Pergamon, Oxford, 1965 and 1966.

56. R. J. Gillespie and R. F. M. White, NMR and stereochemistry, in Progress in Stereochemistry, vol. 3 (edited by P. B. D. de la Mare and W. Klyne), Butterworths, London, 1962, pp. 53-94.

57. A. F. Casy, PMR Spectroscopy in Medicinal and Biological Chemistry, Academic Press, London, 1971.

58. W. B. Jennings, Chem. Rev., 75:307 (1975).

59. G. Binsch, The study of intramolecular rate processes by dynamic nuclear magnetic resonance, in Topics in Stereochemistry, vol. 3 (edited by E. L. Eliel and N. L. Allinger), Wiley, New York, 1968, pp. 97-192.

60. I. O. Sutherland, in Annual Reports on NMR Spectroscopy, vol. 4 (edited by E. F. Mooney), Academic Press, New York, 1971, p. 71.

61. N. S. Bhacca and D. H. Williams, Applications of NMR Spectroscopy in Organic Chemistry, Holden-Day, San Francisco, 1964, pp. 49-54.

62. D. H. Williams and I. Fleming, Spectroscopic Methods in Organic Chemistry, 2d ed., McGraw-Hill, Maidenhead, Berks, 1973, pp. 101-110.

63. W. A. Thomas, NMR spectroscopy in conformational analysis, in Annual Reviews of NMR Spectroscopy, vol. 1 (edited by E. F. Mooney), Academic Press, New York, 1968, pp. 43-89.

64. W. A. Thomas, NMR spectroscopy as an aid in conformational

analysis, in Annual Reports on NMR Spectroscopy, vol. 3 (edited by E. F. Mooney), Academic Press, New York, 1970, pp. 91-147.

65. R. J. Abraham and G. Gatti, J. Chem. Soc. B, 961 (1969).

66. A. F. Cockerill, G. L. O. Davies, R. C. Harden, and D. M. Rackham, Chem. Rev., 73:553 (1973).

67. R. E. Sievers, ed., Nuclear Magnetic Resonance Shift Reagents, Academic Press, New York, 1973.

68. J. D. Roberts, G. E. Hawkes, J. Husar, A. W. Roberts, and D. W. Roberts, Tetrahedron, 30:1833 (1974).

69. I. M. Armitage, L. D. Hall, A. G. Marshall, and L. G. Werbelow, Determination of molecular configuration from lanthanide-induced proton NMR chemical shifts, in Nuclear Magnetic Resonance Shift Reagents (edited by R. E. Sievers), Academic Press, New York, 1973, pp. 313-339.

70. O. Hofer, The lanthanide-induced shift technique: applications in conformational analysis, in Topics in Stereochemistry, vol. 9 (edited by N. L. Allinger and E. L. Eliel), Wiley, New York, 1976, pp. 111-197.

71. C. Kutal, Chiral shift reagents, in Nuclear Magnetic Resonance Shift Reagents (edited by R. E. Sievers), Academic Press, New York, 1973, pp. 87-98.

72. H. L. Goering, J. N. Eikenberry, G. S. Koermer, and C. J. Lattimer, J. Am. Chem. Soc., 96:1493 (1974).

73. M. D. McCreary, D. W. Lewis, D. L. Wernick, and G. M. Whitesides, J. Am. Chem. Soc., 96:1038 (1974).

73a. W. H. Pirkle and D. L. Sikkenga, J. Org. Chem., 42:1370 (1977).

74. J. B. Stothers, Carbon-13 NMR Spectroscopy, Academic Press, London, 1972, pp. 390-433.

75. N. K. Wilson and J. B. Stothers, Stereochemical aspects of ^{13}C-NMR spectroscopy, in Topics in Stereochemistry, vol. 8 (edited by E. L. Eliel and N. L. Allinger), Wiley, New York, 1974, pp. 1-158.

75a. H. Hiemstra and H. Wynberg, Tetrahedron Lett., 2183 (1977).

76. W. R. Jackson and S. Thapebinkarn, Tetrahedron Lett., 1283 (1975).

77. J. C. Speakman, Crystallography and stereochemistry, in Progress in Stereochemistry, vol. 2 (edited by W. Klyne and P. B. D. de la Mare), Butterworths, London, 1958, pp. 1-38.

78. H. H. Mills and J. C. Speakman, Crystallography and stereo-chemistry, in Progress in Stereochemistry, vol. 4 (edited by B. J. Aylett and M. M. Harris), Butterworths, London, 1969, pp. 273-298.

79. G. H. Stout, Lloydia, 35:354 (1972).

80. J. A. Kapecki, J. Chem. Educ., 49:231 (1972).

81. P. J. Wheatley, The Determination of Molecular Structure, 2d ed., Oxford University Press, London, 1969.

82. J. M. Bijvoet, A. F. Peerdeman, and A. J. Van Bommel, Nature, 168:271 (1951).

83. V. W. Laurie, Acc. Chem. Res., 3:331 (1970).

84. E. B. Wilson, Chem. Soc. Rev., 1:293 (1972).

85. E. Wyn-Jones and R. A. Pethrick, The use of ultrasonic absorption and vibrational spectroscopy to determine the energies associated with conformational changes, in Topics in Stereochemistry, vol. 5 (edited by E. L. Eliel and N. L. Allinger), Wiley, New York, 1970, pp. 205-274.

86. M. M. Green, Mass spectrometry and the stereochemistry of organic molecules, in Topics in Stereochemistry, vol. 9 (edited by N. L. Allinger and E. L. Eliel), Wiley, New York, 1976, pp. 35-110.

87. G. C. K. Roberts, The conformation of catecholamines in solution, in Molecular and Quantum Pharmacology (edited by E. Bergmann and B. Pullman), Reidel, Dordrecht, 1974, pp. 77-93.

88. W. G. Richards and C. R. Ganellin, Calculations on unstable con-formations of histamine and methylhistamine, in Molecular and Quantum Pharmacology (edited by E. Bergmann and B. Pullman), Reidel, Dordrecht, 1974, pp. 391-400.

89. A. Golebiewski and A. Parczewski, Chem. Rev., 74:519 (1974).

89a. N. L. Allinger, Calculation of molecular structure and energy by force-field methods, in Advances in Physical Organic Chemistry, vol. 13 (edited by V. Gold and D. Bethell), Academic Press, London, 1976, pp. 1-83.

90. I. G. Csizmadia, Theory and Practice of MO Calculations on Organic Molecules, Elsevier, Amsterdam, 1976.

91. W. G. Richards and M. E. Black, Quantum chemistry in drug research, in Progress in Medicinal Chemistry, vol. 11 (edited by G. P. Ellis and G. B. West), North-Holland, Amsterdam, 1975, pp. 67-90.

92.   J. P. Malrieu, J. Chim. Phys. Phys.-Chim. Biol., 73:319 (1976).

93.   B. Pullman, The adventures of a quantum-chemist in the kingdom of pharmacophores, in Molecular and Quantum Pharmacology (edited by E. Bergmann and B. Pullman), Reidel, Dordrecht, 1974, pp. 9-36.

94.   J. A. Pople, Tetrahedron, 30:1605 (1974).

95.   J. A. Pople, Bull. Soc. Chim. Belg., 85:347 (1976).

96.   A. Pullman and B. Pullman, Quart. Rev. Biophys., 7:505 (1975).

Selected Readings

Stereochemistry

M. S. Newman, Steric Effects in Organic Chemistry, Wiley, New York, 1956.

E. L. Eliel, Stereochemistry of Carbon Compounds, McGraw-Hill, New York, 1962.

J. Grundy, Stereochemistry, The Static Principles, Butterworths, London, 1964.

E. L. Eliel, N. L. Allinger, S. J. Angyal, and G. A. Morrison, Conformational Analysis, Wiley, New York, 1965.

G. Hallas, Organic Stereochemistry, McGraw-Hill, New York, 1965.

K. Mislow, Introduction to Stereochemistry, Benjamin, New York, 1966.

E. L. Eliel, Elements of Stereochemistry, Wiley, New York, 1969.

J. L. Pierre, Principles de Stéréochimie Organique Statique, Armand Colin, Paris, 1971.

J. D. Morrison and H. S. Mosher, Asymmetric Organic Reactions, Prentice-Hall, Englewood Cliffs, New Jersey, 1971.

C. J. Hawkins, Absolute Configuration of Metal Complexes, Wiley, New York, 1971.

W. L. Alworth, Stereochemistry and Its Application in Biochemistry, Wiley, New York, 1972.

G. Natta and M. Farina, Stereochemistry, Longman, London, 1972.

J. E. Fergusson, Stereochemistry and Bonding in Inorganic Chemistry, Prentice-Hall, Englewood Cliffs, New Jersey, 1974.

W. Klyne and J. Buckingham, Atlas of Stereochemistry: Absolute Configuration of Organic Molecules, Chapman and Hall, London, 1974.

F. D. Gunstone, Guidebook to Stereochemistry, Longman, London, 1975.

R. W. Giese, R. P. Mikulak, and O. A. Runquist, Stereochemistry: An Introductory Programmed Text, Burgess Publ. Co., Minneapolis, 1976.

Symmetry and Group Theory

H. H. Jaffé and M. Orchin, Symmetry in Chemistry, Wiley, New York, 1965.

R. M. Hochstrasser, Molecular Aspects of Symmetry, Benjamin, New York, 1966.

F. A. Cotton, Chemical Applications of Group Theory, 2d ed., Wiley, New York, 1971.

A. Nussbaum, Applied Group Theory for Chemists, Physicists and Engineers, Prentice-Hall, Englewood Cliffs, New Jersey, 1971.

J. M. Hollas, Symmetry in Molecules, Chapman and Hall, London, 1972.

J. D. Donaldson and S. D. Ross, Symmetry and Stereochemistry, Wiley, New York, 1972.

C. A. Mead, Symmetry and Chirality, Topics in Current Chemistry, vol. 49, Springer-Verlag, Berlin, 1974.

A. Vincent, Molecular Symmetry and Group Theory, Wiley, London, 1977.

Structural and Electronic Chemistry

Tables of Interatomic Distances and Configuration in Molecules and Ions, The Chemical Society, London, 1958; Supplement, 1956–1959, published 1965.

L. Pauling, The Nature of the Chemical Bond, 3d ed., Cornell Univ. Press, Ithaca, New York, 1960.

C. A. Coulson, Valence, 2d ed., Oxford Univ. Press, London, 1961.

H. B. Gray, Electrons and Chemical Bonding, Benjamin, New York, 1964.

E. Cartmell and G. W. A. Fowles, Valency and Molecular Structure, 3d ed., Butterworths, London, 1966.

M. W. Hanna, Quantum Mechanics in Chemistry, 2d ed., Benjamin, Menlo Park, California, 1969.

D. S. Urch, Orbitals and Symmetry, Penguin, Harmondsworth, England, 1970.

R. J. Gillespie, Molecular Geometry, van Nostrand, London, 1972.

T. E. Peacock, The Electronic Structure of Organic Molecules, in The International Encyclopedia of Physical Chemistry and Chemical Physics, vol. 2, topic 4, Pergamon, Oxford, 1972.

J. Hine, Structural Effects on Equilibria in Organic Chemistry, Wiley, New York, 1975.

N. D. Epiotis, W. R. Cherry, S. Shaik, R. L. Yates, and F. Bernardi, Structural Theory of Organic Chemistry, Springer-Verlag, Berlin, 1977.

Author Index

Numbers in parentheses are reference numbers and indicate that an author's work is referred to although his name is not cited in the text. Underlined numbers give the page on which the complete reference is listed.

Subject Index